Florida's Aquatic Butterfly Gardens

Florida's Aquatic Butterfly Gardens

How to Create a Beautiful Backyard Habitat for Attracting 70+ Species with 100+ Native Plants

SEAN PATTON AND
KENDALL SOUTHWORTH

Palm Beach, Florida

Pineapple Press
An imprint of The Globe Pequot Publishing Group, Inc.
64 South Main Street
Essex, CT 06426
www.globepequot.com

Distributed by NATIONAL BOOK NETWORK
Copyright © 2025 by Sean Patton and Kendall Southworth

British Library Cataloguing in Publication Information available

Library of Congress Cataloging-in-Publication Data

Names: Patton, Sean, 1993– author. | Southworth, Kendall, 1999– author.
Title: Florida's aquatic butterfly gardens : how to create a beautiful backyard habitat for attracting 70+ species with 100+ native plants / Sean Patton and Kendall Southworth.
Other titles: How to create a beautiful backyard habitat for attracting 70+ species with 100+ native plants
Description: Palm Beach, Florida : Pineapple Press, [2025] | Includes index. | Summary: "Kendall Southworth and Sean Patton, specialists in creating aquatic ecosystems for pollinators and wildlife, share their knowledge and skills in this guidebook, showing how aquatic habitats can be used to create the critically important ecosystems necessary for butterfly species of Florida"— Provided by publisher.
Identifiers: LCCN 2024034978 (print) | LCCN 2024034979 (ebook) | ISBN 9781683344643 (paperback) | ISBN 9781683344650 (epub)
Subjects: LCSH: Butterfly gardening—Florida. | Butterfly gardens—Florida. | Water gardens—Florida.
Classification: LCC QL544.6 .P38 2025 (print) | LCC QL544.6 (ebook) | DDC 638/.578909759—dc23/eng/20240904
LC record available at https://lccn.loc.gov/2024034978
LC ebook record available at https://lccn.loc.gov/2024034979

♾️™ The paper used in this publication meets the minimum requirements of American National Standard for Information Sciences—Permanence of Paper for Printed Library Materials, ANSI/NISO Z39.48-1992

CONTENTS

INTRODUCTION

Florida has hundreds of thousands of intricately connected man-made and natural water bodies, forming a living network of ecological relationships. No matter what size or shape they take, these habitats all have one thing in common: they are absolutely vital to the well-being of our communities, human and nonhuman alike. From flood prevention to water filtration and purification, there are countless functions that our aquatic ecosystems perform to keep our environment healthy and resilient.

Many of these ecological functions are not only well known but also hard to ignore. After all, who hasn't seen water levels rise when a thunderstorm rolls in or noted the crawling dune plants protecting our coastlines? Still, there is an aspect of our aquatic ecosystems that goes almost entirely unnoticed and yet is crucial to the survival of hundreds of species—their use as butterfly habitat.

When you envision a butterfly garden, it isn't likely that you'll picture the shoreline of a backyard pond or a tidal marsh. Yet, of the nearly two hundred species of butterfly that call Florida their home, around 30 percent are closely associated with wetlands. This means they rely on aquatic host plants or depend heavily on aquatic nectar sources at various times throughout the year, particularly when others aren't readily available. A host plant is a plant that meets the specific dietary requirements for a butterfly species' caterpillars. Take milkweeds (*Asclepias* spp.), for example. You may have already heard about the importance of planting milkweed for monarch butterflies (*Danaus plexippus*). The reason why milkweed is so essential to the survival and reproduction of the monarch is because it is the only group of plants that hosts its larvae. The monarch caterpillars feed exclusively from the leaves of plants within the genus *Asclepias*, making the conservation and restoration of the various species within the milkweed genus paramount to the continued survival of the monarch. There are several milkweed species native to Florida that grow wild in wet habitats such as cypress swamps, lake and pond margins, and even ditches! These aquatic sites are critical stopovers for monarchs during their migration, due not only to the presence of aquatic host plants but also to an incredibly diverse array of nectar sources. In fact, wetlands are one of the most biologically diverse ecosystems on the planet, on par with rainforests and coral reefs.

Despite their immense value, aquatic ecosystems are often the least understood and most disrupted. Florida has lost more than nine million acres of wetlands since 1845—more than any other state in the nation. While wetlands restoration and public education surrounding the importance of aquatic ecosystems has been slowly increasing over the last few decades, hundreds of thousands of wetlands acres are being drained each year.

The good news? If you're a Florida resident, chances are that you live within a mile's reach of one of the more than seventy-five thousand retention ponds, sumps, channels, wetlands, and lakes that connect our communities' stormwater systems. These bodies of water are often regarded as empty vessels meant to increase property value through a crisp, clean aesthetic of vibrant turf grasses and sparkling blue water. Or they're regarded as purely functional, holding stormwater and preventing flooding like a reservoir. These landscapes are almost always plagued by management issues, whether it be unsightly algae growth or the presence of harmful invasive species. To curb these problems, millions of dollars are spent on herbicides, fertilizers, insecticides, and algaecides. Unfortunately, the frequency and intensity at which these chemicals are applied often results in severe disruptions to the ecosystem. The influx of nutrients into the system encourages the growth of future algae blooms and drops oxygen levels, occasionally resulting in fish kills and, at the very least, harming invertebrates and other smaller organisms. To make matters worse, these applications cause algae to grow increasingly more resistant to the algaecides that are supposed to discourage its growth, a positive feedback loop with negative consequences like chemical dependency for lake management. Worst of all, these polluted waters flow into our ocean, harming the ecosystem along with the indispensable industries of tourism and fishing.

While these traditional management practices are cause for concern, these habitats can easily be transformed into healthy, resilient, and biologically diverse hotspots for pollinators and native wildlife. By utilizing the tools that already exist in nature, you can filter nutrients and pollutants out of the system, prevent shoreline erosion, reduce the frequency and intensity of algae blooms, and improve overall water quality and clarity. By planting an assortment of native plants in and around a water body, you can restore the ecosystem and create a beautiful butterfly garden, all while providing the iconic, tropical appearance Floridians know so well!

BUILDING YOUR BUTTERFLY GARDEN

Simply put, a butterfly garden is any habitat that provides host and pollinator plants for caterpillars and butterflies. Without host plants, a butterfly garden functions only as a visiting site, as they will need to go elsewhere to complete their life cycle. By planting species that serve the insect throughout the entirety of its lifetime, you can construct a hospitable ecosystem that will nourish butterflies and uplift the next generation. Here are some steps you can take to ensure that your butterfly garden is diverse, functional, and resilient.

Make a Plan

Are there specific species of butterfly you would like to see in your landscape? This book lists dozens of butterfly species and their corresponding host plants. If you don't see what you're looking for, do some research and find the host plant that suits the species you desire. Use the information in the section "Growing Aquatic Plants" to determine which plant species are suitable for your landscape and where you might place them. It is often helpful to make a rough sketch so you can have a better picture of what the final product may be. You don't have to be an artist—printing out a photo of the area and marking species with circles works wonders!

Prepare the Area

Before you begin planting, take a look at the land. Are there sprinkler or irrigation pipes that need to be removed, added, or marked first? Can you spot any invasive species like torpedo grass (*Panicum repens*) or Brazilian pepper trees (*Schinus terebinthifolia*) that may hinder the growth of your plants? Remove these and other obstacles, such as large rocks, debris, or turf, prior to planting.

Select Your Plants

As mentioned previously, a host plant is any plant on which a caterpillar can feed. While many host plants also function as pollinator plants, not all pollinator plants host larvae. The ideal butterfly garden will have a wide variety of native flower sizes, shapes, and colors, as many species only frequent certain plants. A variety of nectar sources is key to providing a diverse, strong, and lush habitat. Monarchs, skippers, sulphurs, and smaller butterflies must land on a flower to retrieve nectar. Therefore, thick-stemmed flowers with shallow nectar reserves like swamp sunflower (*Helianthus angustifolius*) are great options for these species. Swallowtails, sphinx moths, and similarly large pollinators like hummingbirds feed midflight, hovering over or beside the flower. This skillset is usually reserved for more tubular flowers with long, thin openings such as coral honeysuckle (*Lonicera sempervirens*) and firebush (*Hamelia patens*). Incorporate these dynamics into your plan to encourage the butterflies you want most to visit.

Plant in Clusters, and Include the Right Structures

Group wildflowers in clusters of three to five, and have at least one multiple of other plant species used on site for pollination. This arrangement helps butterflies spot the swaths of color, reduces overgrazing by larvae, and encourages ecosystem resilience.

The fiery blossom of the scarlet hibiscus (*Hibiscus coccineus*) is a brilliant example of how aquatic wildflowers don't skim on beauty—or size!

Many host plants will also benefit from being planted in different places throughout the garden, as they can be overeaten if clumped too close together in one place. While many equate wildflower and butterfly gardens, it's important to know that many caterpillars seek out the security of shrubs and trees to pupate, and many of these are host plants themselves. Including a variety of trees, shrubs, wildflowers, grasses, and vines is the best way to ensure a diverse, well-resourced crowd of pollinators.

Plant for Every Season

Some species bloom year-round, while others are seasonal bloomers. It is important to plant species strategically in order to ensure that there are plentiful nectar sources throughout the year. While most butterflies are most active in mid- to late summer, it is still crucial to include species that flower in the fall and winter. Butterflies spend the winter in a number of different ways. Some overwinter; others take on migratory journeys; still others survive as eggs, larvae, or pupae. Some are active year-round in specific regions. Planting late season bloomers aids migrating species like monarchs and ensures that food is accessible in a typically diminished period. If you are in an area that does not freeze, year-round bloomers are always great options. However, many short, seasonal bloomers can pack more nectar than other species. A thoughtful, timely balance throughout the garden is needed.

Buy Local and Organic

By purchasing species from local native plant nurseries such as those listed by the Florida Association of Native Nurseries (FANN), you're not only supporting your community but also taking a step toward ensuring the plants you purchase are safe for butterflies. Unfortunately, many plants sold at big box stores that are advertised for butterfly gardens contain pesticides and other chemicals that can harm the organisms. One study found multiple pesticides in every milkweed plant they tested (across fifteen states). In about a third of those plants, they found the residues to be at high enough concentrations to seriously harm monarchs. While not all local nurseries incorporate organic management practices, many that specialize in natives do. Before purchasing, verify that the plants are pesticide-free and safe for your new visitors. Native plants from local sources also tend to be more adapted to your area and local climate, reducing the odds of their being stressed by drought or flooding. And remember, when it comes to buying native and local, don't fret if every leaf isn't perfect. A few bite marks here and there mean the plant is taste-tested—a good sign!

Maintain Your Garden

Native plants often need no fertilizer (aquatic species under the upland zone should receive none) and generally require significantly lower maintenance than traditional landscape plantings. However, some maintenance is always required, mostly in the form of weeding. Keeping a diligent eye open for invasive species will serve you well, as your garden will be most vulnerable to their encroachment immediately after

Amid a stand of alligator flag (*Thalia geniculata*), one of the authors extends her arm to illustrate the grand height of this herbaceous wild-flower. Florida's native plants provide an impossibly varied array of textures, sizes, growth forms, colors, and ecological benefits—from dainty, edible groundcovers to prehistoric-looking, broad foliage.

planting. Of course, hurricanes, rainstorms, and pests may all require your attention. Once the site is established (usually after two to three years), the weeding will slow down substantially and many plants will reach full size. Prune plants as needed, and know that aquatic gardening is exceptionally seasonal. Prepare for explosive growth in the rainy season and much lower maintenance in dryer, cooler months. Over many months of growth, you will already have enjoyed regular butterfly sightings, but at this stage you may want to take a moment to appreciate how far your garden has come and how you have positively contributed to your environment and its wild inhabitants.

GROWING AQUATIC PLANTS

If you're familiar with landscaping in Florida, you've likely heard the phrase "right plant, right place." This refers to the importance of choosing plants that will thrive under the light, moisture, and soil conditions of the location you intend to place them in. This consideration is all the more true for aquatic habitats, which have a variety of additional conditions that change the suitability of a planting location. Not only do aquatic habitats typically have a declining slope leading to water, but they also generally undergo seasonal changes in depth. In fact, it is completely natural for many shallow ponds or swales to lose all standing water during the winter dry season. It is important to become acquainted with these aspects of your aquatic landscape to better understand how the ecosystem functions and how best to build your butterfly garden.

Shoreline Zones

Shoreline depths can be divided into four main zones: upland, riparian, emergent, and littoral. Different regions will have various degrees of this range, with the smallest differences in zones being a few inches in freshwater springs to over 5 feet in some cypress forests! These zones vary immensely in moisture and soil conditions, so by identifying which zones are present in your landscape and which plants call each home, you can optimize your plant choice and location. Different counties and organizations often conflate or define these zones differently. We find it most helpful to define them by how water levels fluctuate throughout the year.

The upland zone is that which is farthest from the water. This area is close to the water but always above the high-water mark, with the exception of intense flooding or extreme weather events. Most trees, large shrubs, and vines occur in this zone, and many of the larger butterflies and moths that host on trees live in this zone, as well as nesting birds that frequent the habitat. An example of an upland zone is the elevated edge of a river or the peak of a dune. This area tends to be the easiest to plant.

Riparian zones are generally between the high-water mark and the edge of the average depth of the water (generally between 5 and 25 centimeters) and are inundated during the rainy season. This is the zone in which you can plant the most varied vegetation, including some of the most aesthetically appealing flowering species, such

This floating butterfly garden was constructed by the authors for the Bay Park Conservancy in Sarasota. The pond in which it rests lies between one of the most frequently trafficked roads of the city and the precious Sarasota Bay, so this feature serves not only as a buffer to help filter stormwater runoff before it reaches the bay but also as a piece of wildlife habitat that provides food and shelter to butterflies, birds, turtles, and underwater creatures! This image shows the island just before its flowering season—imagine how this tiny island transforms when in bloom!

as golden canna (*Canna flaccida*) and scarlet hibiscus (*Hibiscus coccineus*). Typically, the riparian area is inundated only during rainy periods and otherwise is on the dryer side. During these dryer periods, caterpillars move between plants. Most caterpillars occupy this zone, as well as frogs, lizards, nesting ducks, and feeding songbirds. Nearly all swamps, marshes, and estuarine habitats have riparian zones, and bogs, fens, mesic-hydric meadows and grasslands are predominantly riparian habitat.

Emergent zones are underwater most of the year and are exposed only during the dry season, when water levels are at their lowest (generally between 25 centimeters and a meter). This zone has many of the more "traditional" lake planting species, such as pickerelweed (*Pontederia cordata*), duck potato (*Sagittaria lancifolia*), spikerushes, and bulrushes. This area has fewer caterpillars associated with it due to the higher amounts of water, but some species thrive in this zone and pollinator plants are more dominant and easily planted. Several important seed-bearing plants also grow here, which are utilized by bird species. Unfortunately, this area is often improperly constructed in man-made or otherwise disturbed water bodies, with the shoreline dropping directly from the upland to the emergent zone, as opposed to a gradual descent. This dramatically reduces the amount of plantable space, increases erosion, and erases the most prime location for pollinator plants! Before you begin planting, observe the slope of the shoreline to assess what zones are present and the maximum depth your plants can tolerate. The standard 4:1 slope in aquatic management (4 feet across to 1 foot of depth) is an invitation for erosion in the authors' experience. A 6:1 slope is much safer, with some natural slopes in Florida being over 100:1! Cypress and mangrove swamps, sawgrass marshes, and floodplains tend to be emergent habitat.

In many neighborhoods that suffer from eroded shorelines or improperly built slopes, community leaders must find solutions in order to preserve the structural integrity of the properties that surround it. By advocating for restoration methods that establish a properly sloping shoreline, you can be a voice for reintroducing the riparian zone back into the ecosystem. This process not only increases plantable space but also restabilizes sediment and prevents future erosion. Using burlap materials alongside wooden supports and planting vegetation instead of plastic tubes or rocky materials will improve long-term shoreline health and be more cost effective.

Littoral zones are the deepest and most poorly defined zone. Many "littoral shelves" vary wildly in depth and size, ranging from those that are intermittently dry and barely a few centimeters under water to some classified as anything shallower than 2 meters. For the purposes of this book, the authors consider a littoral zone any area with vegetation that is shallower than 2 meters but deeper than a meter during the rainy season and usually exposed only during drought, if ever. This area is too deep for most emergent vegetation to grow and is largely defined by aquatic vegetation that does not break the surface, as well as lily pads. There are almost no host plants located in this area, but it is vitally important for the health of the aquatic ecosystem and contains numerous pollinator plants. Eelgrass (*Vallisneria americana*), fragrant water lily (*Nymphaea odorata*), spatterdock (*Nuphar advena*), and bladderwort (*Utricularia floridana*) are all flowering deep-water species. Animals that frequent littoral areas include diving birds such as anhingas, large fish, turtles, sirens, and manatees in rivers and springs. As defined here, this zone includes ponds, lakes, rivers, and wetlands of relatively advanced depth.

These four zones each have their own flora, fauna, and function. While being very different, they are inextricably connected and support the well-being of the rest of the habitat. An ideal healthy aquatic butterfly garden will make use of all four zones.

Ebbs and Flows

Many people often worry about seeing shallow ponds dry out or seeing shoreline exposed during dryer months. Not only is this process entirely normal for many water bodies, but it is also a crucial time in which many plants reseed and send up new growth, colonizing bare patches of shoreline. By planting more densely upfront, you can encourage thicker stands of growth that will provide coverage even during the fall and winter months. During the dry months, planting groundcovers that tolerate seasonal changes will prevent erosion and provide aesthetic value.

There are many other benefits to these natural ebbings as well, such as the concentration of fish and animals into central shallow pools, which are essential for many migratory birds like spoonbills, wood storks, and herons. This process is timed carefully by many birds, with their migrations, nesting, and hatchlings' first flight timed around the seasons and these ephemeral water bodies.

A wetland or rain garden that has more extensive dry periods is usually a riparian habitat, allowing the gorgeous flowers that dominate this zone to be front and center. Consider the area you intend to plant and how its water levels fluctuate over the course of a year. With this information, you can build a garden that only gets *more* beautiful and resilient as a consequence of seasonal fluctuations.

MAINTAINING A HEALTHY AQUATIC HABITAT

To establish and maintain a functional butterfly habitat for your water body, you must strive for the overall well-being of the entire ecosystem. As previously mentioned, management issues are all too common when it comes to aquatic landscapes in Florida. The main problems are erosion, invasive plants, and algae blooms. All three of these issues can be mitigated by a diverse planting of native species, which doubles as your butterfly garden! Remember, like any landscaping, your new-and-improved habitat will require some maintenance. Typically, these plantings require nothing more than occasional harvesting, and that's only if growth reaches an excess. By planting densely and diversely throughout the full range of pond zones, you can drastically reduce time, effort, and costs spent on maintenance.

Erosion

There are three primary causes of natural erosion: rain and storm runoff, waves, and wildlife. Runoff occurs when there is more water than the land can absorb, with excess liquid flowing into water bodies like ponds and swales. This excess is largely from rainfall but can also come from rooftops, naturally low-lying channels, sprinklers, and anywhere water routinely flows. This runoff often contains high amounts of nutrients from fertilizers used in home landscapes and pollutants from streets and other highly trafficked areas. Having a thick border in the upland zone of a water body not only provides a natural filtration mechanism but also reduces the erosion caused by such flows.

While most water bodies will experience normal levels of wave-based erosion from wind, this effect is often heightened to a more concerning degree due to fountains and aerators being placed too close to shore. This type of erosion forms steep banks at pond margins, often resulting in costly restoration projects down the line. Plantings in the riparian and emergent zones will help this situation, buffering the impact of waves along the perimeter of the body.

Although many assume bulkier species like turtles are to blame for most animal-based erosion, it is actually fish that are far more often than not the offenders. Tilapia and sailfin catfish (commonly called *Plecostomus*) are the worst culprits. These fish form holes and burrows up to 5 feet in length into banks. Both are nonnative and detrimental to the health of a pond, destroying native plants and outcompeting native beneficial organisms. Some native plant species, such as knotted spikerush (*Eleocharis interstincta*), softstem bulrush (*Scirpus tabernaemontani*), or bald cypress (*Taxodium distichum*), have thick, complex root systems that make it quite hard for burrowing to occur. If you know any of these invasive fish species are present in your ecosystem, consider prioritizing planting species that reduce their burrowing and try alternative methods of control, such as manual removal and fishing them out.

One form of erosion that is certainly not natural is mowing. Yes, you read correctly! Since so many man-made water bodies are constructed with a steep slope and turf grass right to the edge, there is little supporting the shoreline. Having heavy mowers routinely cutting grass along the edge of a pond gradually squeezes these banks into the water body. This activity causes erosion and the pond to infill rather rapidly. To combat this issue, try using "no mow" groundcovers or short native grasses, or install a beautiful living shoreline with a margin of several feet.

Invasive Species

Invasive species management is one of the largest and most costly sectors of environmental management. An invasive species is one that has been introduced to an area outside of its natural range (intentionally or otherwise) and disrupts native communities, sometimes to the extent of erasing native plant populations. Invasive species wreak havoc on biodiversity and cause ecosystems to be less dynamic and resilient. However, native species have evolved to the state's unique conditions, enabling many to endure salt, drought, and extreme weather conditions. Native butterflies and other pollinator species have coevolved with these native plant species over time, and these communities depend on one another for reproduction and survival.

Much to environmental restorationists' dismay, many garden centers and nurseries continue to promote and sell invasive species, profiting off the growing interest in supporting declining butterfly populations. Tropical milkweed (*Asclepias curassavica*), one of the most popular offenders, is often the only milkweed species available at such businesses and is advertised for use in butterfly gardens. Its showy blooms, quick regrowth, and easy propagation have made it well known and beloved among homeowners looking to support local pollinators, but many are unaware of the plant's dire effects. Tropical milkweed has been spreading throughout the native ecosystems of central and southern Florida, displacing the twenty-one other native milkweed species native to Florida. Moreover, tropical milkweed becomes a larger problem when planted in temperate areas where it doesn't die off during winter. A protozoan parasite, *Ophryocystis elektroscirrha* (OE), evolved alongside monarch butterflies and travels with those infected, being deposited on leaves when they land. When caterpillars hatch and begin feeding, they ingest the parasite, which replicates inside them. Higher levels of OE have been linked to lower migration success and reductions in flight ability, body mass, and mating success. The parasite-host relationship is usually manageable, as monarchs are able to live well and reproduce while carrying normally occurring levels of OE. Since native species die back after blooming or seasonally, the parasite is killed as well, ensuring a future growing season of parasite-free vegetation. Since tropical milkweed lacks this natural limiting mechanism, higher levels of OE accumulate on the plant and cause successive generations of monarchs to be exposed to fatal levels of OE.

This is only one example of the harms of invasive species. The most common species tend to find their way into your garden thanks to birds, wind, and upstream waters. Some, like torpedo grass, are mixed in with sod. Keeping these invasive species away is sometimes as simple as not using turf grasses and planting native grasses, groundcovers, or mulching instead. Local governments often keep lists of the most harmful invasive species and their control methods, so consider reviewing these to be able to properly identify and manage those that are most common in your area.

Algae Control

In Florida's aquatic ecosystems, algae as we typically refer to it comes in two main forms: filamentous mats (which are thick and hairlike) and planktonic (which is microscopic and turns the water a green, red, or brown hue). Some planktonic algaes are incredibly toxic and are referred to as harmful algae blooms (HABs), the likes of which include the infamous "red tide" along the gulf coast and the blue-green blooms of central Florida. Algae blooms, particularly in spring and summer, are natural and unavoidable in Florida. In fact, algae of all kinds are a crucial aspect of the ecosystem, despite their often unsightly forms. However, when an ecosystem is out of balance, its growth can be excessive—to the extent of endangering humans and wildlife.

One of the main causes of extreme algae growth is an excess of nutrients in the ecosystem. Nutrients found in fertilizers—particularly nitrogen and phosphorus—encourage plant growth and are often used in lawns and gardens. While more organic fertilizers like compost and worm castings will usually not cause issues, synthetic fertilizers (especially liquid fertilizer) are a real threat to water bodies. These fertilizers are not entirely subsumed by the plants at most label rates and instead wash out of yards and into storm drains and nearby lakes, ponds, and wetlands, fueling massive algae blooms in both freshwater and saltwater. A single pound of fertilizer can fuel over a ton of algae in a lake! To prevent this result, never fertilize close to lakes or ponds, and fertilize at least 10 feet away from any paved surface that feeds into a storm drain. If you must use fertilizers, switch to compost, worm castings, or large grain, slow-release fertilizers, which are all less damaging to aquatic habitats.

There are many ways to deal with existing algae blooms. Traditional management practices prioritize chemical applications to kill blooms as they occur. However, this method simply releases nutrients back into the water when the algae dies. These nutrients fuel future algae growth, often now more resistant to the algaecide, requiring more intensive and costly applications in the future. Additionally, many algaecides contain copper, which kills many native plants and wildlife while building up detrimentally over time, never breaking down!

Prevention through nutrient control and native plantings is the best way to tackle algae growth in a future-oriented, cost-effective way. Manual harvesting of algae mats physically removes nutrients from the system and allows the material to be utilized as a natural fertilizer elsewhere. Stocking native algae-eating fish is another option to encourage beneficial nutrient cycling. Installing a floating island will soak up nutrients and shade out algae. Shade trees or lily pads can block light to algae and keep it at tolerable levels. Aerators can be helpful in shallower, stagnant ponds that benefit from destratification. Some natives like coontail (*Ceretophyllum demersum*) or those of the *Chara* genus release chemicals that kill algae. When in doubt, don't spray, and spray only as a last resort. Instead, reach for prevention and harvesting methods that will encourage natural mechanisms of nutrient cycling and balance in the ecosystem.

Get Set to Get Wet!

Now that you're familiar with the basics of aquatic gardening, you're ready to explore the world of aquatic butterfly gardening! Be it sunny or shady, constantly wet or intermittently dry, salty, or sandy, you will find suitable species to bring life and vibrancy to your garden. Throughout this process, you'll find yourself beginning to recognize the diversity of species around you, perhaps alongside an increased feeling of familiarity and connection to the natural world. You'll watch butterflies complete their life cycle, birds harvest seeds and berries from shrubs and wildflowers, and native bees and dragonflies roam. You'll get your hands dirty, an activity proven to boost feel-good chemicals like serotonin that reduce anxiety and improve brain function.

Florida's unique ecosystems are disappearing rapidly, along with populations of native butterflies and wildlife. These habitats form an interconnected, beautiful natural matrix that provides us with drinking water, clean air, flood protection, soil stability, recreational activities, and natural beauty. Don't be surprised if you notice the same species you planted in your garden pop up in nearby habitats—due to the interrelatedness of our landscapes, this is a common and enormously beneficial effect of restoring native habitat.

Take a moment to appreciate your dynamic, lush garden, new perspectives, and the knowledge that the positive environmental impacts of your plantings reach far beyond your garden.

Upland   Riparian   Emergent   Littoral

Stocking Savvy
Freshwater Pond Zones in Florida

High Water Mark

This illustration depicts species that are found in all four zones of a freshwater pond. The authors would like to note the species included are not to scale.

1

TREES

AMERICAN ELM
(Ulmus Americana)

In the nineteenth and twentieth centuries, the deciduous American elm was a go-to for greening developing cities. The fast-growing elm was cherished as an urban specimen, with a delicate but sturdy form that lasted a century or more. They were particularly beloved for forming dreamy archways over streets. Unfortunately, this landscaping style resulted in devastation when a fungal disease caused catastrophic die-offs to the plant, which had developed little resistance to pests and illness. In the wild, American elms don't form "pure" stands, and since they were mostly sourced from only a handful of cultivars, the trees had little genetic diversity, making them extremely susceptible to such fatal events.

The simple, deeply serrated leaves of the American elm emerge from branches that spread with grace and dignity. Gazing on the canopy from afar, it is easy to see why this tree became such a prominent staple of the country's early cities.

The painted lady has the longest migration of any butterfly species in the world. They're also quick fliers, capable of reaching 20 miles per hour!

While the tree is still listed as endangered in North America, Florida has yet to see an occurrence of Dutch elm disease. Growing wild in our mesic forests and hardwood swamps, the American elm is a worthy candidate for any aquatic butterfly garden. The tree hosts the eastern comma (*Polygonia comma*), mourning cloak (*Nymphalis antiopa*), painted lady (*Vanessa cardui*), and red-spotted purple (*Limenitis arthemis*) butterflies, as well as a number of moths. It produces clusters of small, bell-shaped green flowers in early spring, before the reappearance of its leaves. Its seeds are a valuable food source for birds when little else is available.

The winged elm (*Ulmus alata*) is a more drought-tolerant, upland alternative that additionally hosts the question mark butterfly (*Polygonia interrogationis*). This sibling has a shallow root system and can be slightly smaller in stature. It will not grow in more southern counties, like the American elm does, and is also susceptible to Dutch elm disease.

Family: Ulmaceae (Elm Family)
Hardiness: 8A–10A
Lifespan: Long-lived perennial (150+ years)
Zone: Upland
Soil: Occasionally inundated to not extremely dry clay, loamy, or sandy soil

Depth: Shoreline, no permanent inundation
Exposure: Full sun to partial shade
Growth Habit: 60–70 feet tall and 20–40 feet wide
Propagation: Seed, cuttings

ASHES

(Fraxinus spp.*)*

Of the small handful of ashes native to Florida, two stand out as candidates for the aquatic garden: pop ash (*Fraxinus caroliniana*) and green ash (*Fraxinus pensylvanica*). With a name like pop ash, it may come as no surprise that the tree is perhaps best known for its use as a choice material for electric guitar bodies. Its wood is lightweight and easy to sand, and it takes additives like paints and finish well, making it a go-to for the industry. However, the pop ash (also known as swamp ash) is much more than its utility to humans. Native to many of Florida's aquatic landscapes, including floodplains, swamps, and similar sites with prolonged, deep inundation, the pop ash is a prime option for restoration. It is also the larval host of the eastern tiger swallowtail (*Papilio glaucus*), mourning cloak (*Nymphalis antiopa*), and viceroy (*Limenitis archippus*) butterflies.

Pop ash is one of the most flood-tolerant trees in Florida, rivaling oaks and cypresses while retaining a significantly smaller stature. During Hurricane Ian in 2022, record flood waters hit Myakka River State Park, causing 4 feet of flooding in some areas, but the pop ash escaped unscathed! Like many ashes, these trees are favored hosts of American mistletoe (*Phoradendron leucarpum*) and their associated butterfly, the great purple hairstreak (*Atlides halesus*). A single pop ash might provide habitat for dozens of other species, including epiphytic plants, butterflies, and birds.

The green ash is a similarly suitable choice for the aquatic landscape and is also quite adaptable to urban locations. In fact, it is one of the most widely planted

The clustered, hanging paddle-shaped structures of this green ash are winged fruits called samaras.

The orange sulphur is one of the most common butterfly species in North America and is known as an adaptable generalist. Its larva is called the alfalfa caterpillar.

ornamental trees across the nation. Recognized primarily for its interesting foliage, the green ash produces flowers on hanging panicles that lack petals. In addition to hosting the aforementioned butterflies of the pop ash, green ash hosts the orange sulphur (*Colias eurytheme*).

The ash trees of the South are a critical food source for native frogs. Along seasonal ephemeral water bodies, leaf litter provides advantageous feeding opportunities for tadpoles. Many amphibians need temporary, fishless water bodies in which to reproduce. The shallow wetlands and bogs where one may find the green ash are ideal for this purpose. Since Florida has far fewer amphibians than other states in southern Appalachia due to the high connectivity of our waterways, the presence of ashes in these critical water bodies creates an essential opportunity for those present. Small mammals and birds, particularly cardinals, enjoy the ash tree's seeds.

When choosing a plant from a nursery, be sure the specimen has one central trunk and well-spaced branches along its form. If two major branches are opposite one another, remove one to encourage strength and long-term structure. Some nurseries will "top" the tree in its adolescence to encourage bushy growth—however, this is not considered good practice.

Family: Oleaceae (Olive Family)

Hardiness: 8A–10B for *F. caroliniana,* 8A–9B for *F. pensylvanica*

Lifespan: Long-lived perennial (60–100+ years)

Zone: Riparian–Upland

Soil: Aquatic—somewhat moist clay, loamy, sandy, or mucky soil for *F. caroliniana; F. pensylvanica* is less aquatic and tolerant of only occasional inundation

Depth: Shoreline, no permanent inundation; *F. caroliniana* tolerates longer seasonal flooding up to several feet in more extreme floodplains

Exposure: Full sun to partial shade

Growth Habit: 30–60 feet tall and 10 feet wide for *F. caroliniana;* 50–80 feet tall and 30 feet wide for *F. pensylvanica*

Propagation: Seed

The leaves of the bald cypress are quite small and flat, appearing on a structure called a branchlet that gives the plant the false appearance of having a pinnately compound leaf.

BALD CYPRESS

(Taxodium distichum)

From its head to its "knees," the bald cypress commands attention. It is at home not only along water bodies but also within them. It has several peculiar qualities, including allelopathic leaves that prevent weeds from sprouting beneath and the formation of "knees," which are woody projections of roots that extend beyond the ground or water. It was previously believed that this structure provides oxygen to the roots, but it is now more commonly assumed that this trait is related to stabilization and structure. These knees are more common in waterlogged soils or on the downward side of slopes. Regardless of the true reason behind this curiosity, the root system of the cypress provides immense benefits in preventing erosion and stabilizing sediment. In fact, bald cypress is often the top pick for shoreline stabilization, not only due to its underground resources but also by keeping mowers away from shorelines, where the equipment can push banks in with their weight over time.

The feathery foliage of the bald cypress is another distinguishing feature of the tree, which sheds its leaves in winter following the autumnal period, when its needle-like leaves turn a golden rust. It is a valuable roosting and nesting site for birds, which enjoy its seeds along with small mammals.

> **Fun Fact:** In 2012, scuba divers discovered an underwater bald cypress forest off the coast of Mobile, Alabama, with trees that cannot be dated with radio-carbon methods. This suggests the trees could be more than fifty thousand years old and therefore likely lived in the early glacial interval of the last ice age.

Under the right conditions, the bald cypress has a lifespan exceeding a thousand years, with one specimen in North Carolina being deemed the ninth-oldest tree in the world, with a tree-ring count dating back to 605 BC.

For a more compact, drought-tolerant option, consider the pond cypress (*Taxodium ascendens*). While this relative can still reach heights of 80 feet, it's comparatively smaller and tends to put up fewer "knees," which makes these trees a good option for planting farther up the shoreline in the home landscape.

Family: Cupressaceae (Cypress Family)
Hardiness: 8A–10B
Lifespan: Long-lived perennial (400+ years)
Zone: Littoral–Upland
Soil: Aquatic to somewhat moist clay, muck, loamy, or sandy soils
Depth: Depending on their original size, this species can be grown in littoral zones or as an emergent; smaller trees should be started on shorelines or riparian habitat
Exposure: Full sun to partial shade
Growth Habit: 50–75 feet tall and 30–45 feet wide
Propagation: Seed

While it is rare to hear the flower clusters of the box elder referred to as "showy," it's hard to say they're not at least quite interesting.

BOX ELDER
(Acer negundo)

Box elder is a rather unappreciated native tree of Florida. While it is not known to have strong wood or an exceptionally long lifespan for a tree (typically living around sixty years), its ecological value, in tandem with a hardy adaptability, is quite impressive. Found naturally in floodplains, along streams, and colonizing disturbed sites, it is tolerant of both occasional inundation and extremely dry, short periods. The tree will grow almost anywhere without salty wind or spray, providing a 40–50-foot-tall rounded canopy in which an assortment of birds and wildlife may enjoy its shelter, seeds, fruit, and flowers. The latter arrive early in the warm months and are yellow in color, emerging in clusters from pinkish racemes. While these blooms are not particularly showy, they make up for it by attracting a wide assortment of pollinators. The box elder is also the host plant for the largest moth native to North America, the cecropia moth (*Hyalophora cecropia*).

The tree's lack of popularity can be attributed to its weak wood and damage-prone branches, qualities that are common for similarly fast-growing but short-lived trees. These traits mean that box elder should not be planted in locations where broken or fallen limbs could be a hazard to property or individuals. However, it is highly recommended for use in landscapes where other trees would not be expected to do well. Since it is a pioneer of open, aquatic or semi-aquatic sites, it does wonders for stabilizing moist soils and preventing erosion with its fibrous root systems and prolific seeding habit. After its death, nesting mammals and birds will call it home, and its decaying wood will provide food and shelter for many species of fungi and insects that will serve as food for other organisms.

Family: Sapindaceae (Soapberry Family)

Hardiness: Zones 8A–9B

Lifespan: Long-lived perennial (60–100 years)

Zone: Riparian–Upland

Soil: Moist to inundated sandy, loamy, or clay soils, and tolerant of extremely dry periods for a short time

Depth: Shoreline, no permanent inundation

Exposure: Full sun to partial shade

Growth Habit: 40–50 feet tall and wide

Propagation: Seed

Male banded hairstreaks are known for being fierce defenders of territory and will engage in extended aerial fights. It is common to see individuals with chunks of wing missing for this reason.

Buttonwood is distinguishable from other species thanks to its Dr. Seuss–like, petalless flower clusters. While one might be inclined to think these adorable features are its namesake, "button" actually refers to its fruits, which look like old, leathery buttons.

BUTTONWOOD

(Conocarpus erectus)

Growing wild alongside mangroves (and sometimes referred to as the "fourth mangrove" for this reason), buttonwood is a staple of Florida's shorelines. It is one of the more commonly utilized natives in landscaping, as it is easily pruned into a small shrub, screen, or hedge. When allowed to grow in open space alongside salty winds, the tree will often form a sculptural, contorted appearance that can be quite enchanting. Native to areas with tidal inundation, the tree is clearly tolerant of salty, sandy coastal conditions, but it will still retain its adaptability in sites farther inland. Its intricate root system helps stabilize shorelines and prevent erosion, and its foliage provides shelter and nesting sites for birds, insects, and marine life.

The buttonwood is the host plant for the martial scrub hairstreak (*Strymon martialis*) and is a choice nectar plant for many species of bees and butterflies, which enjoy its green flowers that form compact heads on panicles. The cultivar *Conocarpus erectus* var. *sericeus* provides an attractive, silvery appearance that stands out brightly in a mixed landscape planting. Like many plants with silver foliage, it has a slightly hardier drought resistance and is smaller than its green counterpart. Air plants (*Tillandsia* spp.), saw palmetto (*Serenoa repens*), and sea lavender (*Heliotropium gnaphalodes*) all have this distinctive coloration, serving as an example of convergent evolution in which several unrelated plant species display similar characteristics.

Family: Combretaceae (White Mangrove Family)

Hardiness: 9B–11

Lifespan: Long-lived perennial (50–150 years)

Zone: Riparian–Upland

Soil: Regularly inundated to very dry sandy soils and limerock

Depth: Shoreline, no permanent inundation, tolerates saltwater inundation

Exposure: Full sun to partial shade

Growth Habit: 20–40 feet tall and wide

Propagation: Seed (stratified)

The martial scrub hairstreak is a rare find along the southern coasts of Florida.

A stand of cabbage palms resembles a guild of slender, wise elders gazing on their verdant realm. It is no wonder this palm has woven itself so deeply into the Southeast's art and culture.

CABBAGE PALM

(Sabal palmetto)

The state tree of Florida, the cabbage palm is best known for its iconic silhouette of fan-shaped fronds that form a rounded canopy standing tall among the coastal plains, marshes, and hammocks throughout the entirety of our state all the way to the Carolinas. It is a pioneer species, with innumerable benefits to pollinators, wildlife, and humans alike. Early Floridians made use of all parts of the palm for food, medicine, weaving, and shelter. Today, the cabbage palm is an enduring symbol of our natural heritage.

In late spring, the palm produces groups of small, white blossoms that develop into clusters of dark purple berries, which serve as a valuable source of food for wildlife. Its dead fronds, which hang at the base of the canopy, are crucial roosting habitat for bats, whose fur color blends in seamlessly with the leaflets. These fronds are also choice habitat for tree frogs! The "frizz" from the leaves provides excellent nesting material, and the palm boots are important and unique habitat for other species, such as the golden polypody fern. The cabbage palm is the larval host for the monk skipper butterfly (*Asbolis capucinus*). In fact, so many species rely on cabbage palms that they are sometimes referred to as the "tree of life." For an in-depth view of this tree, refer to Jono Miller's recently published *The Palmetto Book*.

One of the largest skippers of North America, the monk skipper claims the name capucinus due to its resemblance in color to the vestments of Capuchin monks.

Unfortunately, many landscaping companies remove the green fronds in a practice called "hurricane cutting," leaving the trees with an overpruned mohawk. This process takes cherished habitat away from our native wildlife and can seriously harm the tree, reducing its nutrients and opening it up to diseases (actually reducing its ability to survive severe weather events like hurricanes in the process). The dead leaves will be shed on their own, and in the meantime, leave the pruning behind and allow the cabbage palm to display its full, timeless beauty.

Sabal minor, or dwarf palmetto, is an alternative shrubby *Sabal* for the landscape with similar pollinator benefits and a restricted growth habit. Dwarf palmetto displays classic fan-shaped fronds, and it enjoys relatively shaded, moist areas, although it is fairly drought tolerant once established.

Family: Arecaceae (Palm Family)
Hardiness: 8A–11
Lifespan: Long-lived perennial (100+ years)
Zone: Riparian–Upland
Soil: Occasionally inundated to very dry loam and sand

Depth: Shoreline, no permanent inundation, tolerates longer seasonal flooding
Exposure: Full sun to full shade
Growth Habit: Up to 100 feet tall and 10–20 feet wide
Propagation: Seed

The early spring bloom period of the Carolina willow is much appreciated by the multitude of insects that frequent its catkins.

CAROLINA WILLOW

(Salix caroliniana)

Found wild in every county in Florida, the Carolina willow is a graceful and common feature of pond shorelines, swamps, marshes, and floodplains. While this species enjoys having wet feet, it can grow well in upland zones and tolerates root disturbance well. In early spring, either before or along with its leaves, the willow produces clusters of inconspicuous white blooms that precede elongated catkins that are quite attractive. These flowers are greatly enjoyed by pollinators, particularly the viceroy (*Limenitis archippus*), mourning cloak (*Nymphalis antiopa*), and red-spotted purple (*Limenitis arthemis*) butterflies, who rely on the willow for hosting their larvae.

Without the presence of fire, *S. caroliniana* can transform herbaceous wetlands into forested wetlands, an evolution often prevented by prescribed fires. One of the very few trees that can grow in perpetually flooded areas, it is a true littoral species, able to grow and thrive in inundation of several feet. Often needing to start in shallower areas, it will gradually creep into emergent and eventually deep littoral zones, with some notable specimens having been found in water over 6 feet deep during the rainy season!

Fun Fact: The Carolina willow is fed to giraffes, antelope, elephants, and rhinoceri at Disney's Animal Kingdom Park, thanks to its high dry-matter content of protein, fiber, vitamins, and minerals.

Family: Salicaceae (Willow Family)

Hardiness: 8A–11

Lifespan: Perennial (data is limited, but we estimate 30–50 years based on average lifespans of similar *Salix* spp.)

Zone: Littoral–Upland

Soil: Aquatic to not extremely dry sand or muck

Depth: If started in riparian or shoreline areas, can grow several feet deep into littoral depths

Exposure: Full sun to full shade

Growth Habit: 25–60 feet tall and 20–40 feet wide

Propagation: Seed, cuttings

It's a monarch! No, wait . . . It's a viceroy! Confused? Good—it's working! The viceroy is a Müllerian mimic of the monarch. Müllerian mimicry is a theoretical phenomenon in which two or more species with effective defenses (such as noxiousness) share a similar appearance. The viceroy and monarch share extremely similar markings, with only very subtle wing differences. Both species are noxious to predators, and therefore both benefit from having a similar appearance.

The growth habit of the Florida strangler fig is almost mythical in appearance. It's hard to deny its eccentric allure, even though its parasitic habit can be quite destructive contextually.

FLORIDA STRANGLER FIG
(Ficus aurea)

The Florida strangler fig is truly one of the more strange and intriguing trees native to our state. Like all trees, it begins its life as a seed, but its germination usually takes place in the canopy of another tree. The infant tree spends its adolescence as an epiphyte, shooting out aerial roots as it develops that reach hungrily for the ground to begin their sprawling, parasitic spread. As it grows, it strangles the tree from which it came, becoming freestanding and quite large under the right conditions. For this reason, it's recommended to avoid planting these species in smaller landscapes or in areas where the aesthetics of the fig's draping, sinuous roots would not be welcome. Those willing to embrace the wild nature of the fig will be rewarded in its immense ecological value and one-of-a-kind spectacle.

Like all figs, the tree has a mutualistic relationship with fig wasps, who are its sole pollinators. These species are a brilliant example of coevolution, with each depending on the other for survival and reproduction. The flowers of the fig are fully encompassed by a syconia, accessible only from the outside via the ostiole. Within *Ficus aurea*, both male and female flowers are contained within this structure. Female wasps, drawn by the volatile attractant released by the female flowers, squeeze through the ostiole. Once inside, they pollinate the flowers, lay their eggs, and die. In a little over a month, adult wasps emerge within the syconia. The males mate with the females, cutting exit holes through the walls of the fig, ready to enter the world. When the female wasps

While the aptly named Antillean daggerwing is found in the West Indies, it strays to the Florida Keys, making it a special find if you're lucky enough to spot it.

emerge later, the male flowers of the fig have just matured, and the females pack their bodies with pollen before leaving through the helpful holes created by their mates. Over the following week, the fruit ripens and is consumed by birds and small mammals, who disperse the seeds. This intricate dance is one of the most well understood and revered displays of symbiosis in the natural kingdom.

Although the fig is well known for its partnership with the fig wasp, it is also the host for the ruddy daggerwing (*Marpesia petreus*) and Antillean daggerwing (*Marpesia eleuchea*) butterflies.

Family: Moraceae (Mulberry Family)
Hardiness: 9B–11
Lifespan: Long-lived perennial (100+ years)
Zone: Upland
Soil: Moist to occasionally inundated—dry sand, muck, and limerock

Depth: Shoreline, no permanent inundation, very salt tolerant
Exposure: Partial shade to full shade
Growth Habit: 40–60 feet tall and 20–60 feet wide
Propagation: Seed, cuttings

The hackberry was one of the most common plant remains found at the Meadowcroft Rock Shelter, the oldest site of human habitation of North America, suggesting hunter-gatherers made use of the hackberry as far back as sixteen thousand years ago.

HACKBERRY

(Celtis laevigata)

It's hard to find a home or urban landscape the hackberry wouldn't excel in. While it grows naturally in floodplains and along the banks of streams, it is moderately tolerant of heat and drought once established as well. The adaptability doesn't end there—it is also wind, salt, and pollution tolerant, transplants easily, can grow in full sun or partial shade, and grows rapidly.

The hackberry is also host to a large number of butterflies: the hackberry emperor (*Asterocampa celtis*), mourning cloak (*Nymphalis antiopa*), tawny emperor (*Asterocampa clyton*), and question mark (*Polygonia interrogationis*), as well as the American snout (*Libytheana carineta*), for which it is the sole larval host. Its green flowers appear in early spring, and its cherished fruits are eaten by a wide number of birds and small mammals.

The delightful taste and nutrition of the fruit is not known only to woodland creatures—humans also have an ancient, extensive culinary and medicinal history with the hackberry. Many indigenous American communities have used the berries for jellies and cakes, dried them to form a spice, and made them into decoctions to treat sore throats and regulate menses. Hackberries are a valuable source of dietary fiber, vitamins, protein, and pigments.

Family: Ulmaceae (Elm Family)
Hardiness: 8A–10B
Lifespan: Long-lived perennial (150+ years)
Zone: Riparian–Upland
Soil: Consistently wet to not extremely dry clay, loam, and sand

Depth: Shoreline, no permanent inundation
Exposure: Full sun to partial shade
Growth Habit: 50–70 feet tall and wide
Propagation: Seed

You might be inclined to think the most interesting thing about the hackberry emperor is that it relies solely on the hackberry to host its larvae. You would be wrong. The hackberry emperor has a variety of unique behaviors that have made it known as a rather odd, mercurial insect. It flies quickly and erratically, rarely visiting flowers to nectar, instead favoring hackberry sap, feces, and dead animals. It is also known to land on humans, enjoying our sodium-rich sweat.

If you think these spines are bad, you should see the trunk!

HERCULES' CLUB

(Zanthoxylum clava-herculis)

Of all the trees to accidentally graze, Hercules' club is at the bottom of the list. Its name is warning enough—the trunk of the tree *and* its branches are covered in formidable thorns. One encounter is enough to cement the identification in a naturalist's mind, as there is simply no other tree with such imposing qualities. Despite the cautionary tales, its ingenious protective measures shouldn't deter you from incorporating it into your landscape. Its dense foliage provides habitat and food sources for birds, insects, and small mammals. While some believe the early spring blossoms aren't anything to write home about, setting your eyes on a decently sized cluster of the tree's white flowers may prove otherwise. Regardless, these flowers get the job done, attracting pollinators and transforming into berries that are enjoyed by birds and small mammals. It is often used by those who are familiar with the plant as a form of oral anesthetic, as the leaves produce a numbing sensation when chewed. This unique form of stimulation can make a delightful, interactive addition to a sensory garden.

The giant swallowtail is the largest butterfly in North America. Its hungry larva is seen as quite pesky on citrus farms, earning it the name orange dog.

The Hercules' club is a host for the giant swallowtail butterfly (*Papilio cresphontes*). Giant swallowtails host only on citrus species, with hoptree (*Ptelea trifoliata*), sea torchwood (*Amyris elemifera*), and wild lime (*Zanthoxylum fagara*) being other Florida natives that likewise serve as hosts. These species also act as the host plants of the critically endangered island or Schaus' swallowtail (*Papilio aristodemus*), which is found only in the very south of Florida and several islands in the West Indies. Overdevelopment and logging have removed many of the tropical hardwood hammocks in these areas, endangering the dozens of species that call this special ecosystem their home.

Family: Rutaceae (Citrus Family)

Hardiness: 8A–10B

Lifespan: Long-lived perennial (data on life expectancy is limited, but most sources claim it is long lived; similar species live at least fifty years, and it is reasonable to make this assumption for the Hercules' club as well)

Zone: Upland

Soil: Occasionally inundated to very dry clay, loamy, or sandy soils

Depth: Shoreline, no permanent inundation, very salt tolerant

Exposure: Full sun to partial shade

Growth Habit: 10–25 feet tall and 10–25 feet wide

Propagation: Seed, cuttings

While these Jamaican caper flowers are white, the blossoms will turn purple the next day!

JAMAICAN CAPER

(Quadrella jamaicensis, formerly *Capparis cynophallophora)*

With their bold, star-shaped petals and purple stamens, the white blossoms of the Jamaican caper are as eye-catching for humans as they are for pollinators. These flowers bloom in the spring and summer, turning pinkish purple within a few hours of opening, and are particularly attractive to species that are more active in the evenings. Its beige seed pods are another distinctive feature, splitting to reveal a vibrant reddish orange interior. It is the larval host for the Florida white butterfly (*Appias drusilla*).

Hurricane resistant and drought tolerant, the Jamaican caper is found growing wild in coastal hammocks and disturbed sites. It can withstand extremely long, dry periods but is also tolerant of occasional saltwater inundation, such as from storm surge. Its foliage provides significant cover for birds and wildlife, which enjoy the fruit as well. The flowers have extremely long, delicate stamens, leading some to refer to them as "cat whiskers." Jamaican caper is an unusual pop of color in the typically dry and often difficult-to-plant coastal ecosystem.

Family: Capparaceae (Caper Family)

Hardiness: 10A–11

Lifespan: Long-lived perennial (most sources claim this, although specific data is limited; it is safe to say this tree will live at least 20–30 years in appropriate conditions)

Zone: Upland

Soil: Occasionally inundated—extremely dry humus, limerock, or sandy soil

Depth: Shoreline, no permanent inundation, very salt tolerant

Exposure: Full sun to partial shade

Growth Habit: 6–12 feet tall and 6–10 feet wide

Propagation: Seed (scarified)

These Florida whites are displaying a behavior called "puddling," a drinking activity in which they can consume hundreds of gutloads of water, retaining the valuable sodium and minerals and excreting the rest. In many species, this behavior is restricted to males, with studies suggesting the nutrients they collect are then stored in the sperm and transferred to the female during mating as a kind of "nuptial gift."

There's more to the delightful scent of the magnolia flower than you might think. Methyl dihydrojasmonate, a compound found in both the fragrant magnolia and jasmine, is released from the flower and binds to the VN1R1 receptor in the nose, sending signals to parts of the brain that control the limbic system, which is associated with emotions, memory, and motivation.

MAGNOLIAS

(Magnolia spp.)

There are few flowers as aromatic and alluring as those of the magnolia. These natural jewels make the tree one of the more commonly planted species native to Florida in home landscapes, as well as a symbol of Southern elegance and heritage. Boasting broad ecological value, hurricane resistance, and showy fruits and foliage, the magnolia is an exceptional staple of any habitat in which it is placed. The two most popular magnolia species also happen to be the most adapted to aquatic landscapes. The sweetbay magnolia (*Magnolia virginiana*) is the more water tolerant of the pair and is found wild in many types of wetland ecosystems throughout the state, forming clonal thickets in wetter sites. Although it prefers locations of dappled shade in its native habitat, it survives in a diversity of environments. Its adaptability makes it ideal for restoration settings, where it provides a lovely sight along the borders of ponds and swales.

While the size and color of the eastern tiger swallowtail is enough to swoon over, it is the caterpillar that possesses the most remarkable of the species' qualities. Its false eyespots are glaring to predators, and, if touched, a pair of vivid orange glands (called osmeterium) emerge from its neck region like devil's horns, secreting a foul-smelling acidic substance.

The sweetbay magnolia hosts the eastern tiger swallowtail (*Papilio glaucus*). However, as one of the first flowering plants to evolve, it is still primarily pollinated by beetles and does not have true nectar but protein-rich pollen instead.

Southern magnolia (*Magnolia grandiflora*) is a much more popular species worldwide and is often considered one of the most prized ornamentally for its flowers and wood. It is larger than the sweetbay magnolia and more tolerant of drier conditions, as it is native to upland hardwood forests and similar habitats.

Due to the low dispersion range (beetles are poor pollinators and the seed is quite heavy), magnolias are prone to speciation and endemism to relatively small areas. This trait is also seen by a wide number of pawpaws (*Asiminas* spp.) in Florida, such as the Manasota pawpaw (*Asimina manasota*), which is now found only in Manatee County, as its already tiny range combined with increased development has led to its likely extirpation in Sarasota County.

Deer enjoy browsing its leaves and twigs, and its seeds and fruits are consumed by a significant number of bird species and small mammals.

Family: Magnoliaceae (Magnolia Family)

Hardiness: 8A–10B for *M. virginiana* and 8A–9B for *M. grandiflora*

Lifespan: Long-lived perennial (80–120+ years)

Zone: Riparian–Upland

Soil: Wet, somewhat moist sandy or mucky soil for *M. virginiana*; *M. grandiflora* is more tolerant of dry conditions

Depth: Shoreline, no permanent inundation; *M. virginiana* tolerates longer seasonal flooding

Exposure: Partial shade to full shade for best results for *M. virginiana*; *M. grandiflora* is more tolerant of full sun, but both will do well

Growth Habit: 20–30 feet tall and 10–15 feet wide

Propagation: Seed (stratified)

While the musclewood is most prized for its brawny trunk, its serrated leaves can put on quite the canopy display when mature.

MUSCLEWOOD

(Carpinus caroliniana)

Named for its smooth bark texture and ribbing, which occurs over time, musclewood is a slow-growing tree with extremely hard wood, earning it the alternative name "ironwood." It is native to floodplains, river banks, and similar aquatic habitats in Florida. Its notable texture and foliage alone make it a prime choice for planting in the urban or home landscape, where it can also serve as a wonderfully sturdy and flexed structure for climbing if you have the muscles for it!

While it is found wild as an understory tree in low, shady places, it displays an impressive tolerance of sunnier, drier locations. During droughts in southern counties, having some irrigation will be an aid. Musclewood will tolerate pruning for use as a hedge or screening plant, but it is arguably most handsome when allowed to form multiple trunks and low branches. Many nurseries sell single-stemmed trees as an option as well, which is ideal for more urban settings. Its comfort in shade combined with its shorter stature makes it a perfect fit for locations where space and sun may be limiting factors.

Additionally referred to as American hornbeam, musclewood is a standout feature of any butterfly garden not only because of its desirability for a variety of pollinating insects but also for its use as a larval host for the eastern tiger swallowtail (*Papilio glaucus*), striped hairstreak (*Satyrium liparops*), and red-spotted purple (*Limenitis arthemis*) butterflies. Its yellow-green flowers are organized in catkins that appear in mid- to late spring. Its foliage, which is occasionally browsed by deer, turns a delightful faint yellow-orange-red in the fall.

Family: Betulaceae (Birch Family)
Hardiness: 8A–9B
Lifespan: Long-lived perennial (50–100+ years)
Zone: Riparian–Upland
Soil: Well-drained to inundated sandy, loamy, or clay soils, and moderately tolerant of drought

Depth: Shoreline, no permanent inundation, salt tolerant
Exposure: Partial shade to full shade, although occasionally tolerant of full sun
Growth Habit: 20–30 feet tall and wide
Propagation: Seed (stratified)

Can you count how many species have taken up residence in this oak? Neither can we! Oaks can provide a home for literally hundreds of plant and wildlife species.

OAKS

(Quercus spp.*)*

When one pictures the Southern landscape beyond the coast, it is the oak that comes to mind. Oaks, hardwood trees of the *Quercus* genus, are the majestic guardians of the varied ecosystems they inhabit, providing food and shelter for hundreds of creatures. With their resilience, longevity, and keystone ecological role, their importance cannot be overstated. Some oaks tower impressively over the lands they rule, while others stand humble and shrubby.

There are two main categories of oak: white and red. Florida has nineteen native oak species that fall in these two categories. White oaks are sometimes referred to as "annual oaks" because their acorns mature in one growing season. These acorns have fewer tannins and are much sweeter in taste. They are more common for use in construction and furniture making and are generally associated with mature forests, where they reveal subdued fall colors of rust, yellow, and brown. Red oaks, by contrast, drop their more bitter acorns after two years and tend to have bark that is darker in appearance, with deeper ridges and furrows. Red oaks play similar ecological roles but are more associated with disturbed sites and variable habitats, like early succession forests. These often display more vibrant fall colors as well, such as red, orange, and dark purple.

For the aquatic butterfly garden, there are a handful of species to choose from, including the laurel oak (*Quercus laurifolia*), water oak (*Quercus nigra*), swamp chestnut oak (*Quercus michauxii*), and the most popular—the live oak (*Quercus virginiana*). These oaks all host the Horace's duskywing (*Erynnis horatius*) and the white-M (*Parrhasius m-album*) butterflies. The laurel and water oaks additionally host the red-banded hairstreak (*Calycopis cecrops*), and the swamp chestnut oak the gray hairstreak (*Strymon melinus*). Live oaks, with their rough bark, long lifespans, and massive size, are a staple for dozens of epiphytic plants, with hundreds of species airplants, orchids, and ferns relying on them. It is also likely that some species of hairstreak butterflies utilize the various mistletoe species that parasitize oaks as larval hosts, but the mistletoe may be conflated with the oaks themselves. The authors would like to note that studying arboreal-preferring butterfly species is more difficult than studying those that prefer shorter, more herbaceous plants, due to the inaccessibility of observation.

Family: Fagaceae (Beech Family)

Hardiness: Species are found 8A–11, with distribution being species dependent

Lifespan: Long-lived perennial (generally 100+ years, with species variation; for example, *Q. nigra* typically lives 30–50 years)

Zone: Upland

Soil: Species dependent

Depth: Species dependent; *Q. nigra*, *Q. laurifolia*, *Q. virginiana*, and *Q. michauxii* are the most water tolerant and seasonal flood tolerant. Local ecotypes may be able to tolerate near year-round inundation, but all generally need occasional periods of dryness. Riparian habitat is uncommon but observed in many *Quercus* species, especially along rivers. *Quercus* species do not do well with all roots inundated year-round and will not survive being planted underwater or placed completely in the water. Plant them always as shoreline or upland species.

Exposure: Species dependent

Growth Habit: Species dependent

Propagation: Seed

The petite flowers of the persimmon transform into sweet fruits that many claim to be in a taste category of their own. However, consuming the astringent, unripe fruit will leave your mouth numb or chalky due to the high tannin content, which fades with maturation.

PERSIMMON

(Diospypros virginiana)

The wild persimmon has been a fundamental aspect of the Southern landscape for centuries. Indigenous Floridians used all parts of the tree for everything from treating fever to playing dice-like games. During the Civil War, citizens used the seeds for buttons and, when blockades limited access to coffee, for their morning cup. With the introduction of the nonnative Japanese persimmon, the cultural familiarity and interest in planting the delightful and ecologically valuable wild persimmon faded. Still, our native persimmon has much to offer, even beyond its prized fruits.

Found commonly along wetland borders where soil stays moist but not regularly inundated, persimmon is a fast-growing tree that often "volunteers" in disturbed and drier sites. While it can reach sizable heights in the wild, it keeps a relatively short stature in home and urban landscapes. Persimmons should be planted in locations where fruit and leaf litter are not disturbances, nor the assortment of mammals and birds that will come to reap their treasures.

Persimmon is a host plant to several moth species, most notably the luna moth *(Actias luna)*. Since luna moths do not drink nectar after hitting adulthood, planting a hackberry or persimmon is often the only way to guarantee their presence in a given area.

The persimmon's small, bell-shaped greenish yellow blossoms, which appear in late spring to early summer, are pollinator attractors. The plant is dioecious, so it's important to plant both male and female trees to ensure successful fruiting. If you're not sure which you have, you can note, which male flowers appear in threes, while female flowers emerge in solitude.

Family: Ebenaceae (Persimmon or Ebony Family)

Hardiness: 8A–10B

Lifespan: Long-lived perennial (60–100+ years)

Zone: Upland

Soil: Moist to occasionally inundated—dry clay, loam, or sand

Depth: Shoreline, no permanent inundation

Exposure: Full sun to partial shade

Growth Habit: 35–60 feet tall and 15–35 feet wide

Propagation: Seed

Swamp bay is a relative of the avocado (*Persea americana*), which has an understandably similar flower.

SWAMP BAY

(Persea palustris)

A slender, evergreen tree, swamp bay has a simple and elegant form. It is found naturally in swamps, wet flatwoods, and similar wetland habitats. The top side of its leaves is green and glossy, while the underside is pale and silvery, with an orange tinge along the midvein that separates it from relatives. When crushed, its foliage delivers a distinct aroma.

The swamp bay is a larval host for the palamedes swallowtail (*Papilio palamedes*) and spicebush swallowtail (*Papilio troilus*) butterflies, as well as a nectar source for many other pollinators. Its seeds are eaten by birds. Unfortunately, members of the Laurel family are susceptible to the fatal laurel wilt disease, so be cautious when purchasing to ensure that your chosen specimen is entirely healthy.

Florida is also home to two other *Persea* species, with red bay (*Persea borbonia*) being another species right at home in wetland borders, as well as the endemic silk bay (*Persea humilis*).

Family: Lauraceae (Laurel Family)
Hardiness: 8B–10B
Lifespan: Long-lived perennial (data is limited, but it is safe to say swamp bay can live to at least 50 years in the appropriate site conditions)
Zone: Riparian–Upland

Soil: Occasionally inundated to somewhat moist muck or sand
Depth: Shoreline, no permanent inundation, tolerates longer seasonal flooding
Exposure: Full sun to partial shade
Growth Habit: 15–30 feet tall
Propagation: Seed

The palamedes swallowtail, which is quite elegant in flight, has shown to decrease in number in areas that have been heavily impacted by laurel wilt disease—another reason to engage in management practices that limit its spread.

In bloom, the mature swamp dogwood can put on an impressive display with its white flower clusters.

SWAMP DOGWOOD
(Cornus foemina)

Rightfully named, swamp dogwood is a resident of Florida's swamps and similar areas that experience moderately brief and shallow inundation. It is an ideal option for planting along retention ponds or streams, either as a specimen tree or as a screen. The tree is classically attractive, with dark green, heavily venated leaves and showy clusters of white flowers that appear in late spring. Once pollinated, these blooms transform into beautiful blue berries, which are eaten by birds, squirrels, and other wildlife. Deer and rabbits will occasionally browse its foliage and twigs.

Swamp dogwood is the host to the spring azure butterfly (*Celastrina ladon*) and attracts all sorts of pollinating insects. The more popularly planted flowering dogwood (*Cornus florida*) is an alternative for upland locations that still retain plenty of moisture. It is incredibly showy and one of the most beloved flowering trees native to Florida. Historically, the bark from various species from the *Cornus* genus was used by Civil War soldiers to treat pain and fever, and a poultice made from the leaves was utilized for open wounds. Today, its wood is prized by artisans for making everything from arrows to jewelry boxes.

Family: Cornaceae (Dogwood Family)

Hardiness: 8A–10A

Lifespan: Long-lived perennial (<50 years)

Zone: Riparian–Upland

Soil: Wet to moist sandy, loamy, or mucky soils

Depth: Shoreline, no permanent inundation, tolerates longer seasonal flooding

Exposure: Full sun to partial shade

Growth Habit: 10–30 feet tall and 10–20 feet wide

Propagation: Seed

Fascinatingly, the larvae of the spring azure butterfly are tended to and cared for by ants! The caterpillars produce a nectar-like secretion that not only feeds the ants but also prevents them from causing harm. In return, the ants will defend the caterpillars from predators.

From above, the leaves of the water hickory appear vibrant and glossy. Standing beneath the canopy, however, the pale undersides of the compound leaves flicker in the softest of breezes, creating an oscillating scene of colors as different sides of the leaves are exposed.

WATER HICKORY

(Carya aquatica)

A primary species in many floodplains, water hickory plays a pivotal role in water filtration and erosion control. While its nut and leaf drop can become a nuisance in urban or home landscape settings, it is a great choice for restoring habitat in landscapes where the plant has the space to provide its resources to a wide variety of wildlife. The tallest of the hickories, this tree reaches outstanding heights (up to 100 feet) when provided open space and favorable conditions, greatly enjoying the wet soils of Florida's inland aquatic ecosystems. Water hickories are also known to grow quite old, often well over a century, making them good canopy trees near buildings. Like all larger trees, be sure to place it 10 feet or more from the dwelling.

One of water hickory's most distinguishing traits is its peculiar foliage—growing to 15 inches long, its pinnately compounded leaflets feature a dark green, glossy top and pale bottom. Its relatively inconspicuous greenish blooms flower in early spring, attracting butterflies and other pollinating insects. Also referred to as swamp hickory or bitter pecan, water hickory is a host for several moth species, including the famed luna moth (*Actias luna*). For shadier upland locations, choose its sister, the pignut hickory (*Carya glabra*). It is similar in physical appearance, nut and leaf spread, soil preferences, and pollinator benefits but is far more drought tolerant and grows wild in mixed upland forests. It is one of the more common hickories of the South and is a historically important source of timber, forming wagon wheels and skis alike.

Family: Juglandaceae (Walnut or Hickory Family)

Hardiness: 8A–9B

Lifespan: Long-lived perennial (150+ years)

Zone: Riparian–Upland

Soil: Well-drained to inundated sandy, loamy, or mucky soils, and moderately tolerant of drought

Depth: Shoreline, no permanent inundation, tolerates longer seasonal flooding

Exposure: Full sun to partial shade

Growth Habit: 30–80 feet tall by 25–50 feet wide

Propagation: Seed (stratified)

Like many plants, the multitrunked wax myrtle can appear as a small, stately tree or a dense, branching shrub depending on pruning and site conditions.

WAX MYRTLE

(Morella cerifera)

Native to all counties in Florida, the evergreen wax myrtle is an exceptionally hardy and versatile plant for the home landscape. It is tolerant of a wide range of soil, sun, salt, and moisture conditions, making it suitable for a broad variety of landscape settings. Historically, the plant has been prized in the South for its medicinal and practical uses—in fact, the plant is also referred to as candleberry, as its waxy berries were used by early American settlers for candle making. Today, its leaves are used in culinary settings, as an antibacterial herbal remedy, and to make essential oils for use in aromatherapy.

The wax myrtle is equally relied on by nonhuman creatures. Its fruits are a valued food source for many birds and small mammals, particularly during the winter. It provides shelter and nesting opportunities as well, and it is the larval host for both the banded hairstreak (*Satyrium calanus*) and the red-banded hairstreak (*Calycopis cecrops*) butterflies.

Through a symbiotic relationship with specialized bacteria, wax myrtle converts atmospheric nitrogen gas (N_2) into a nitrogen form readily absorbed and utilized by plants. While legumes are known among gardeners for this specific nitrogen-fixing property, most don't realize the wax myrtle is even more efficient at this process. This is a great quality to have as one of the first plants to colonize an area.

The red-banded hairstreak tends to prefer feeding on the fallen, decaying leaves of the wax myrtle, as opposed to the live plant.

In the landscape setting, this species will perform as well as part of a rain garden as it will an upland hedge. It tends to sprout from its roots, which may not be desirable aesthetically for some homeowners. Despite this potential issue, consider keeping this growth for its bonus environmental value. Like many plants, the wax myrtle can be considered a tall shrub or relatively small tree depending on pruning and site conditions.

Family: Myricaceae (Bayberry Family)

Hardiness: 8A–11

Lifespan: Perennial (individual specimens will rarely live past 50 years and the plant is therefore usually described as short lived, but its suckering nature ensures its persistence far past this)

Zone: Riparian–Upland

Soil: Wet—somewhat moist sandy or mucky soil

Depth: Shoreline, no permanent inundation, tolerates longer seasonal flooding

Exposure: Full sun to partial shade

Growth Habit: 10–15 feet tall and 8 feet wide

Propagation: Seed, cuttings

The unforgettable pod of the West Indian mahogany makes this species a tough one to forget once you've seen it in the wild.

WEST INDIAN MAHOGANY

(Swietenia mahagoni)

A South Florida native, the West Indian mahogany is best known for its use in the lumber and furniture industries due to its rich color, straight grain, and wood durability. Unfortunately, due to overharvesting, the species is now listed as "Threatened" in the Preservation of Native Flora of Florida Act and is protected by state, federal, and international laws.

The mahogany is a sizable, robust tree with a rounded canopy—a structure and appearance perfect for planting in parks, along streets and medians, and on the margins of ponds and lakes. Found naturally in our tropical rockland and coastal hammocks, the tree attracts a truly diverse community of pollinators, despite having fairly inconspicuous flower clusters. The West Indian mahogany is also host to the only other mistletoe species native to Florida, the critically endangered mahogany mistletoe (*Phoradendron rubrum*)! This species can reestablish a secure population only via the mass restoration of mahogany in South Florida and the Caribbean. There are also unconfirmed reports that there is a unique subspecies of the great purple hairstreak (*Atlides halesus*) that may have coevolved with the mahogany mistletoe in its isolated island and extreme South Florida habitats. Further research and restoration can help us learn more about and preserve these species.

Family: Meliaceae (Mahogany Family)

Hardiness: 10A–11

Lifespan: Long-lived perennial (100+ years)

Zone: Upland

Soil: Somewhat moist—short, very dry periods of clay, limerock, loamy, or sandy soils

Depth: Shoreline, no permanent inundation, very salt tolerant

Exposure: Full sun

Growth Habit: 30–70 feet tall and 40–60 feet wide

Propagation: Seed

The white fringe tree takes the old saying "good things don't last forever" to heart. We doubt the verity of that axiom, but we certainly miss the blooms when they're gone!

WHITE FRINGE TREE
(Chionanthus virginica)

If you're looking for a visually stunning native tree for your landscape, look no further than the white fringe tree. In full bloom, the tree looks nearly snowfallen, blossoming with thousands of pure white, slightly fragrant flowers that hang in long, delicate panicles. While these blooms last only for around two weeks in spring, their famed beauty makes incorporating this tree into your landscape worth it. While the cultivation of this species is still relatively fringe, its prevalence in the trade is expanding, and it is locally available in many counties.

The fringe tree blooms best in full sun locations, but its foliage appears more attractive with a scattering of afternoon shade. They make exquisite understory trees and can create a dazzling display when planted in a mixed garden location where they receive morning sun. Commonly found in upland woods and along stream banks, the tree prefers moist, acidic soil and will tolerate wetter soils as well. Its fruits are eaten by birds and small mammals, and it is the larval host for many moth species, particularly sphinxes.

Family: Oleaceae (Olive Family)

Hardiness: 8A–9B

Lifespan: Long-lived perennial (up to ~50 years)

Zone: Riparian–Upland

Soil: Well-drained to occasionally wet sandy, loamy, and clay soils

Depth: Shoreline, no permanent inundation

Exposure: Full sun to partial shade

Growth Habit: 10–20 feet tall and 8–15 feet wide

Propagation: Seed (stratified)

2
SHRUBS

BAY CEDAR
(Suriana maritima)

The now-endangered bay cedar used to be a staple of our dunes, beaches, and similar coastal habitats. In fact, it's endemic to our state! Incorporating it into the landscape setting is a great way to increase the survival and reproduction of this species, and gardeners will be glad they did. The evergreen shrub reveals its dainty yellow flowers year-round, and its gray-green, succulent foliage is unlike any other. Unsurprisingly, it is highly salt tolerant and provides valuable erosion control benefits. Bay cedar also forms a fine hedge row and has many use similarities to sea lavender (*Heliotropium gnaphalodes*), though the wildlife they attract are different.

Suriana is a monotypic genus, meaning that bay cedar is the only species within it. Between its resilient growth habit and ecological benefits, bay cedar is truly one of a kind.

The bay cedar is the host plant for the martial scrub hairstreak (*Strymon martialis*) and mallow scrub hairstreak (*Strymon istapa*) butterflies. It is also a choice nectar plant for many more species of butterflies. Planting coastal nectaring species such as this one is vital to the survival and reproduction of pollinators that, like many Floridians, prefer to frequent areas where you can taste the salt on the breeze.

The mallow scrub hairstreak prefers nectaring at small flowers, which makes sense considering its dainty size.

Family: Surianaceae
Hardiness: Zones 9B–11
Lifespan: Long-lived perennial
Zone: Upland
Soil: Not extremely dry to very dry sand
Depth: Shoreline, no permanent inundation, very salt tolerant and can tolerate coastal inundation from storms
Exposure: Full sun to partial shade
Growth Habit: 6–10 feet tall and wide
Propagation: Seed

The whimsical flower promises the inviting appearance of the tasty berry.

BLUEBERRIES

(Vaccinium spp.*)*

While Florida may be known for its oranges, its blueberries aren't far behind! Our state is home to at least eight native blueberry species, and some sources suggest we may have been the first ever to try our hand at commercial blueberry cultivation over a century ago. At the grocery store, you're most likely to find the highbush (*Vaccinium corymbosum*) or rabbiteye (*Vaccinium ashei*), two species that have been cultivated over time to produce sweet, large berries ideal for wide-scale, packaged consumption. These species thrive in full sun and acidic soils, growing wide and high (up to 12 feet) throughout Florida. Like other species in the *Vaccinium* genus, their foliage is standout, changing color vibrantly as the seasons progress. The reddish green leaves of spring turn blueish green in the summer. The distinctive display continues as the foliage transforms once more to the reds, yellows, oranges, and purples of autumn. Florida lacks the fall complexion of more northern states, but the blueberry remains a welcome exception.

For the Floridian who may not be able to accommodate the far-spreading habit of the highbush and rabbiteye, meet two other blueberry species ideal for the landscape setting: Darrow's blueberry (*Vaccinium darrowii*) and shiny blueberry (*Vaccinium myrsinites*). Found in the pine forests and flatwoods of Florida, these species are significantly smaller, maxing out around 2 feet. Their new growth displays surreal shades of purple and blue alongside older, greener foliage, creating an eccentric look that is unique to the species. The fairytale, bell-shaped flowers are replaced by fruits that are smaller than the commercial varieties but just as delicious. Darrow's blueberry does well in particularly moist sites, while the shiny blueberry is more tolerant of upland locations.

Family: Ericaceae (Heath Family)

Hardiness: *V. myrsinites* grows in zones 8A–10B; *V. darrowii* grows as far south as 10A

Lifespan: Long-lived perennial

Zone: Upland

Soil: Acidic sands, with some variability

Depth: Shoreline, no permanent inundation, although highbush blueberry tolerates seasonal freshwater inundation

Exposure: *V. myrsinites* and *V. darrowii* both enjoy full sun to partial shade

Growth Habit: *V. darrowii* is around 2 feet tall and wide, and *V. myrsinites* tends to grow slightly smaller in height

Propagation: Seed, softwood cuttings, divisions

While the buttonbush is typically white in color, this is an example of a regional color variation in which the bloom, early in its life, puts on an attractive pink display.

BUTTONBUSH

(Cephalanthus occidentalis)

One guaranteed way to start a "pollinator party" in your yard is by planting buttonbush. The globular white flowers are a precious favorite of hummingbirds, bees, butterflies, moths, and other pollinating insects. Pollinators aren't the only ones who enjoy the resources provided by the shrub—deer will also readily browse the foliage, and several species of birds enjoy snacking on the seed. Buttonbush is therefore a staple among Florida native gardeners who appreciate its abundance of nectar, classic glossy foliage, and long lifespan. The unusual shape and prolific number of showy spherical clusters make buttonbush a notable seasonal highlight. There are also some populations in south central Florida that have pink highlights throughout their blooming period, though usually this effect is seen only when first opening. This is a potential avenue for future cultivated varieties with alternate pink coloration. Changes in flower color, as well as similar changes such as dwarf variants and weeping types, are not harmful to pollinators. Double-petaling flowers and denser flower clusters have been shown to occasionally impact pollinator use in cultivated plants.

Often confused with the swarthy skipper, there is little data on the deemed inconspicuous neamathla skipper's particular traits or behaviors—other than it looks cute nestled within a flower!

In southern counties, buttonbush flowers in the summer, but it will bloom throughout both spring and summer in central and northern locations. Without pruning, the bush grows quite round and will sprawl several feet, taking up a considerable amount of space along a pond shoreline or home. It should be planted in locations where the soil retains moderate amounts of moisture, as it grows wild in swamps and marshes. If planted in a drier location, moderate irrigation will be necessary to keep the plant thriving. If you have horses, cattle, or other livestock, be warned that buttonbush is poisonous if eaten in large amounts.

Family: Rubiaceae (Madder Family)	**Depth:** 6 inches or less
Hardiness: Zones 8A–11	**Exposure:** Full sun to partial shade
Lifespan: Long-lived perennial	**Growth Habit:** 5–20 feet tall and 4–8 feet wide
Zone: Riparian	
Soil: Consistently wet to usually moist clay, loam, muck, or sand	**Propagation:** Seed or hardwood cuttings

Cocoplums make great jams and jellies—not only for their flavor but also thanks to their natural abundance!

COCOPLUM

(Chrysobalanus icaco)

The appropriately tropical-sounding cocoplum does it all—depending on the ecotype and cultivar, it can serve as a fruit tree, groundcover, hedge, or shrub screen! The cocoplum is guaranteed to have a growth habit that serves your vision, and it does all of this while being drought and moderately salt tolerant once established. These plants tolerate pruning and general shaping quite well but do not topiary, preferring instead to show off their prominent, broad leaves in their mature form.

The cocoplum has two ecotypes: coastal and inland. The coastal variety is shorter and more sprawling than its inland counterparts, typically not exceeding 6 feet tall. Its fruits, which make great jams and jellies, appear in pink, white, and purple, transforming as they ripen or remaining one color depending on the cultivar. If purchasing from a nursery, the word "horizontal" in the name will distinguish this ecotype. While the inland ecotype tolerates pruning well, it will grow much larger if allowed. Its cultivars are "red tip" and "green tip," named for the color of the new leaf growth. The fruits of the red tip are purple, and those of the green tip can be purple or white.

All cocoplums have glossy, rounded leaves that make the shrub easy to identify from afar. They grow wild in coastal swamps, beaches, everglade tree islands, and along the edges of canals or ditches. The small white flowers attract pollinators, and the fruits are consumed by a wide variety of wildlife. While it is unable to survive a hard freeze, southern Florida gardeners will delight in the versatility and hardiness of this shrub. Both the large, almond-like seed and the grape-tasting fruit are edible and part of a long legacy of Florida foraging.

Family: Chrysobalanaceae (Cocoplum Family)

Hardiness: Zones 9B–11

Lifespan: Long-lived perennial

Zone: Riparian–Upland

Soil: Occasionally inundated to very dry clay or sand

Depth: Shoreline, no permanent inundation, tolerates seasonal flooding and brackish tidal influences

Exposure: Full sun to partial shade

Growth Habit: 10–30 feet tall and 10–20 feet wide

Propagation: Seed or hardwood cuttings

Once you learn to recognize the elderberry, you won't believe how often you'll spot this plant growing in abundance along the highways of our state!

ELDERBERRY

(Sambucus nigra sub. canadensis)

Growing wild along disturbed wetlands, the elderberry is easily overlooked as a roadside weed. For those who know what they're seeing, this plant tells a different, far more valuable story—and even those who don't will find it hard to look away from its spectacle in the early summer, when the plant becomes utterly encompassed in clusters of fragrant white flowers. The elderberry is a fast-growing shrub that serves as choice food for deer, birds, and small mammals—the latter two groups also valuing its use as shelter and nesting habitat. While it can grow quite "leggy," pruning can enhance its form as a small tree or denser, shrubby bush. It spreads via rhizomes and can efficiently self-seed, so plant in locations where these traits are welcome.

Female spicebush swallowtails often drum their forelegs, which possess chemoreceptors, toward a leaf to identify which leaf to deposit their eggs on, assessing the leaf's chemical makeup in order to determine the acceptability of that location.

The fruits and flowers of the elderberry are edible and medicinal when properly prepared. In fact, Hippocrates called the elder his "medicine chest," and it is one of the most commonly used medicinal plants in the world. Most grocery stores sell elderberry syrup, tea, or supplements, touting its antioxidants and potentially immune system–boosting effects. However, the fruits and flowers should not be consumed raw, and only after thorough cooking, as all parts of the plant are poisonous otherwise. The elderberry is also toxic to livestock and should not be grown around unsupervised animals, though poisonings are rare since the taste is poor when unprepared.

Elderberry naturally colonizes drainage ditches and grows well in swamp and wetland borders. Use this vigorous species to block out invasive species, alongside other aggressive native species like Carolina willows (*Salix caroliniana*) or golden cannas (*Canna flaccida*).

Family: Adoxaceae (Moschatel Family)
Hardiness: Zones 8A–10B
Lifespan: Long-lived perennial
Zone: Riparian–Upland
Soil: Consistently wet to not extremely dry clay, loam, or sand

Depth: Can grow in riparian habitat in several inches of water for prolonged periods, but does best in areas higher up the shoreline
Exposure: Partial shade to full shade
Growth Habit: 10–15 feet tall and as wide
Propagation: Seed, cuttings

The striking purple display of the false indigo bush, contrasting with vivid orange, is in a league of its own.

FALSE INDIGO BUSH
(Amorpha fruticosa)

This deciduous, multistemmed shrub is known for its spike-shaped raceme of purple petals and contrasting golden stamens. For this reason, it is sometimes cultivated merely ornamentally, although gardeners will be happy to hear of its multifold ecological value. It is the larval host to a handful of butterflies, including the clouded sulphur (*Colias philodice*), gray hairstreak (*Strymon melinus*), hoary edge (*Cecropterus lyciades*), silver-spotted skipper (*Epargyreus clarus*), and southern dogface (*Zerene cesonia*). Beyond its value as a host plant, it also benefits many other pollinators, including hummingbirds.

The false indigo bush is found wild along pond and stream edges, hardwood forests, and both moist and dry hammocks. Its seeds are eaten by birds, and its foliage is composed of interesting, hair leaflets. This bush would do well as a hedge, screen, or accent plant, pairing nicely with shorter bushes that can cover the occasionally leggy bases while the fantastic flowers put on a topward display, starting orange and fading to the notorious indigo with age. Like all legumes, this species is a nitrogen fixer. Space legumes throughout the garden when possible to see healthier species overall, as well as the wide array of butterflies that prefer them.

Family: Fabaceae (Legume Family)

Hardiness: Zones 8A–10A

Lifespan: Long-lived perennial

Zone: Upland

Soil: Occasionally inundated to dry sand

Depth: Shoreline, no permanent inundation, tolerates brief inundation at the height of wet season

Exposure: Full sun to partial shade

Growth Habit: 6–12 feet tall and 6–10 feet wide

Propagation: Seed, cuttings

While the southern dogface barely opens its wings when resting, you might be lucky enough still to catch a glimpse of its namesake. The black spot on its forewings is supposed to be a dog's eye, with the contrasting yellow and black eliciting a canine's profile.

The fetterbush has a balletic charm, with its rows of bell-shaped blossoms nodding gently down its length.

FETTERBUSH

(Lyonia lucida)

Looking at the flowers of the fetterbush, gardeners will be pleased to know this low-maintenance, evergreen shrub is native to all counties in Florida. Its dainty, bell-shaped blooms range in color from white to pinkish red and drape romantically from elongated inflorescences in the late winter and early spring. They are favored by bees, and many scrub caterpillars hide their pupae and chrysalises in these dense shrubs.

Also known as shiny lyonia, this native has extended rhizomes that often sprout new plants, forming a colony. While it grows fine in dry settings, it is found wild in shrubby bogs and other periodically flooded habitats. Fetterbush does particularly well in creekside sandy locations in central Florida, where much of the year dry, well-drained sands prevent more water-preferring plants from growing but seasonal rains may flood out other species.

There are four other *Lyonia* species native to Florida: *L. ferruginea*, *L. fruiticosa*, *L. ligustrina* var. *foliosifloria*, and *L. mariana*, all of which provide use to pollinators.

Family: Ericaceae (Heath Family)
Hardiness: Zones 8A–10B
Lifespan: Long-lived perennial
Zone: Riparian–Upland
Soil: Occasionally inundated to very dry sand

Depth: Shoreline, no permanent inundation, tolerates mild seasonal flooding
Exposure: Full sun to partial shade
Growth Habit: 2–6 feet tall and 4 feet wide
Propagation: Seed

The red berries and shiny emerald foliage of our native hollies have become as archetypal in the landscape as they have in our culture.

HOLLIES

(Ilex spp.*)*

The three common hollies native to Florida share more than their classic Christmassy look of glossy green leaves and vibrant red berries. The American, yaupon, and dahoon hollies are relatively low maintenance and dependable. With different growing habits and site tolerances, a holly can find its place in nearly any landscape.

Of the three, the dahoon holly (*Ilex cassine*) is the most suitable for aquatic environments, growing wild in swamps and along the edges of ponds. With light to moderate irrigation, the tree can be trained to tolerate seasonally drier soils once established. Like the others, it is suitable for full sun to full shade locations, although its crown and growth habit will be more developed in full sun. Their long-lasting display of berries is gorgeous, usually lasting well into the *holly-days!*

The American holly (*Ilex opaca*) grows in a dense, pyramidal form that is ideal for more traditional, ornamental landscapes. It tolerates full sun to full shade and wet to well-draining soils of various types. This holly showcases whitish bark, a distinguishing feature from the others.

The yaupon holly (*Ilex vomitoria*) comes in many cultivars, including a beautiful weeping variety. It is one of the most resilient hollies and often found in urban locations like sidewalk cutouts and similar sites with limited exposed soil. Like the American holly, it also grows in full sun to full shade and can even tolerate salt and drought. Interestingly, yaupon holly has the highest caffeine tolerance of any plant species native to North America. The presence of caffeine kills off many species of insects, so no

Male ruddy daggerwings can be territorial and have been known to swoop down from trees toward orange or red objects, thinking they've spotted another male.

caterpillars specialize in eating hollies, but their medicinal, edible, and structural uses in a garden are extremely useful, and their blooms still provide nectar to a variety of butterflies and other pollinators. The yaupon holly is the most shrubby and dense of the three species, with dahoon most commonly found in a small tree form.

Family: Myrtaceae (Myrtle Family)

Hardiness: *I. vomitoria* and *I. opaca* grow in 8A–10A; *I. cassine* extends to 10B

Lifespan: Long-lived perennial

Zone: Riparian for *I. cassine*; others enjoy locations farther upland

Soil: Species dependent, most often sand or loam

Depth: *I. cassine* can grow in several-inch-deep water for long periods, but all listed hollies do best in shoreline areas, on edges of wetlands, or with wet feet

Exposure: Species dependent, with *I. cassine* enjoying full sun to full shade

Growth Habit: Species dependent; *I. vomitoria* is the smallest listed species, growing generally 8–25 feet tall and 5–8 feet wide

Propagation: Seed takes 2–3 years, but clones of shoots work well

The nectar of the saltbush is as plentiful as its seemingly infinite flowers.

SALTBUSH

(Baccharis halimifolia)

When not in bloom, the saltbush is an easily overlooked native. Not that it's unattractive —its growth habit and foliage are simply typical of many other species of shrub and not particularly striking. However, for several weeks in the fall, the saltbush turns heads with its white, cotton-like plumes of blossoms. This hazy flush attracts many pollinators, and the wind-dispersed seeds are eaten by small birds and mammals, which utilize its form for cover.

Saltbush is native to our inland and coastal wetlands and can also be found dotting ditches and roadsides throughout the state, oftentimes alongside wax myrtle. As its name reveals, it is highly tolerant of salt, making it an apt choice for seaside landscapes. Its autumn blooming period puts on a show in a season not known for its vibrancy, and while it is deciduous in northern Florida, it remains evergreen in the central and south. It is safe to say saltbush is underutilized in home and urban settings. Although its prolific reseeding and occasionally leggy appearance may not be desirable for all, those willing to embrace its natural habit will find much to appreciate in its aesthetic and ecological benefits.

A very nice mixed hedgerow for moist coastal areas is a combination of saltbush and sea oxeye daisy (*Borrichia frutescens*) at its base. This arrangement helps form a showy bright silver green border where the leggy base of the saltbush is blocked by the sprawling growth form of the oxeye daisy. This is perfect for when conditions are too moist for mangrove or saltmarsh habitat but too dry for dune habitat.

Family: Asteraceae (Sunflower Family)

Hardiness: Zones 8A–11

Lifespan: Long-lived perennial

Zone: Riparian–Upland

Soil: Occasionally inundated to not extremely dry loam or sand

Depth: Shoreline, no permanent inundation, tolerates longer seasonal flooding, very salt tolerant and can tolerate brackish water inundation

Exposure: Full sun

Growth Habit: 7–15 feet tall and 5–7 feet wide

Propagation: Seed

The many alternative common names for the sea oxeye daisy, including beach carnation, sea marigold, and seaside tansy, all referencing the plant's lovely yellow flowers.

SEA OXEYE DAISY

(Borrichia frutescens)

The sea oxeye daisy is a low-growing shrub known for its ability to adapt to a wide range of soil and moisture conditions. Growing throughout the state everywhere from saltmarshes to dry forests, the plant has tropical, succulent-like foliage from which delightful buttons of yellow flowers emerge. These flowers bloom year-round but are particularly numerous in the spring and summer, attracting many pollinating insects. The shrub is a larval host for the great southern white (*Ascia monuste*), gulf fritillary (*Agraulis vanillae*), large orange sulphur (*Phoebis agarithe*), and southern broken dash (*Wallengrenia otho*) butterflies.

Evergreen and salt tolerant, the sea oxeye daisy is a beautifully flowering species along a shoreline or farther upland in the home landscape, where it will form patches. It does best in full sun but will grow in partial sun or dappled shade, where it will just flower less. It acts as a fine base to the roots of leggier, salt-tolerant species and makes a great partner for mangroves, saltbushes (*Baccharis halimifolia*), buttonwoods (*Conocarpus erectus*), and seagrapes (*Coccoloba uvifera*).

Family: Asteraceae (Sunflower Family)
Hardiness: Zones 8B–11
Lifespan: Long-lived perennial
Zone: Riparian–Upland
Soil: Consistently wet to somewhat moist clay, loam, muck, sand, etc.

Depth: Shoreline, no permanent inundation, tolerates daily saltwater inundation during king tide flooding (especially in mangrove borders)
Exposure: Full sun to partial shade
Growth Habit: 2–4 feet tall and 2–3 feet wide
Propagation: Seed, cuttings, division

Curiously, the southern broken dash caterpillars, once they've hatched, construct small nests of grass or leaf pieces tied with silk and will often cover themselves with a piece of foliage when emerging to feed.

The white flowers of the shiny wild coffee emerge daintily from attractive, glossy leaves.

SHINY WILD COFFEE

(Psychotria nervosa)

While the name is misleading, the shiny wild coffee does produce small red fruits that do resemble true coffee beans. These fruits are indeed edible and have been experimentally used to make uncaffeinated brews, but ethnobotanist Dan Smith claimed it shouldn't be considered a coffee substitute anytime soon, saying it resulted in only "bad taste and terrible headaches"—a result of the mild chemical that mimics the sometimes anxiety-inducing effects of caffeine. The shiny wild coffee might not make it to your mug, but it should certainly be considered as an addition to your landscape.

Flourishing in the shade, its leaves are prominently venated, giving the foliage an interesting textured look. The shrub is small and evergreen and does well as an accent understory plant, catching eyes with its bright berries and quaint form. Its delicate white flowers are attractive to a particularly wide variety of pollinators, especially rare butterflies like Schaus' swallowtail or the atala.

While not host to any butterflies, *Pyrausta tyralis* (the coffee-loving moth) is a small specialist in this group with a ruddy red color and yellow spots. The caterpillars are very shy, hiding in curled leaves on the coffees and leaping out if you unfurl the leaf, avoiding detection at all costs. The plant should not be placed in locations that receive more than filtered sunlight and can tolerate dry periods only for a short time. Should a more drought-tolerant species be needed, Bahama coffee (*Psychotria ligustrifolia*) will work and tolerates more sun exposure. Velvet-leaved coffee (*Psychotria tenuifolia*) is often found alongside shiny coffee but is less useful for landscaping purposes, as it is quite leggy and smaller in most cases (on occasion, it will take up a more appealing bushy habit).

Family: Rubiaceae (Madder Family)
Hardiness: Zones 9A–11
Lifespan: Long-lived perennial
Zone: Riparian–Upland
Soil: Occasionally inundated to dry loam, limerock, or sand

Depth: Shoreline, no permanent inundation, tolerates longer seasonal flooding
Exposure: Partial shade to full shade
Growth Habit: 2–6 feet tall and 1–3 feet wide
Propagation: Seed

The rare Schaus' swallowtail was nearly wiped out after Hurricane Andrew in 1992, which left only seventy-three documented individual survivors. The destruction of its tropical hardwood hammock habitat combined with the fact that it produces only one generation per year contributes to its endangerment. Captive rearing, however, continues to prove a successful method of population maintenance and growth.

The white stopper putting out its flowers directly off the branch, a sight not too commonly seen.

STOPPERS

(Eugenia spp. and *Myrcianthes fragrans)*

Simpson's stopper is not a true stopper of the *Eugenia* genus, but it is in the same tribe and functions similarly in the landscape. Many Florida botanists consider it *the* ideal compact tree or shrub. When crushed, its foliage exudes a sweet perfume, and the plant is especially low maintenance. Its beautiful ivory flowers blossom in the springtime and are followed by red berries that are prized by birds, which also prefer its canopy for nesting and protection. The bark is beauteous and the leaves evergreen, so no matter the season or growth form, it provides reliable charm. The gardener is also sure to appreciate its drought and wind tolerance.

The white stopper (*Eugenia axillaris*) is a large, evergreen shrub with a classic beauty. Native to our coastal hammocks, it provides shelter for wildlife and attracts pollinators with an abundance of delicate white flowers that bloom directly off the branches. Its drought tolerance and ability to flourish with intensive pruning makes it a reliable and resilient hedge, screen, or accent plant that would look equally beautiful in formal settings as it would in wilder locations. It has drupe-like berries that are enjoyed by birds and other creatures, even humans! Curiously, the white stopper's fragrance is quite divisive, with some enjoying its aroma and others referring to it as skunk-like, so consider testing your own olfactory response before determining whether this quality is make-or-break to you.

All stoppers provide seasonal pollination in butterfly gardens, as well as higher cover for caterpillars to pupate in. Many *Eugenia* species are found exclusively in southern counties, preferring tropical climates and partial shade. These differ subtly and are less commonly found at nurseries than the Simpson's or white stoppers. However, if you happen to come across a red berry stopper (*Eugenia confusa*), Spanish stopper (*Eugenia foetida*), or red stopper (*Eugenia rhombea*), consider contributing to local biodiversity by adding it to your garden.

Family: Myrtaceae (Myrtle Family)
Hardiness: Generally, zones 9A–11
Lifespan: Long-lived perennial
Zone: Upland
Soil: Occasionally inundated to dry humus, limerock, or sand

Depth: Shoreline, no permanent inundation
Exposure: Full sun to partial shade
Growth Habit: 10–15 feet tall for Simpson's and white, others may be taller
Propagation: De-pulped seed

The masses of cream-colored flowers of *Viburnum* species draw in a diverse crowd of pollinators.

VIBURNUMS

(Viburnum spp.)

Walking along a neighborhood street or city avenue, your chances of spotting the prolific ivory flowers and handsome leafage of viburnums are relatively high, and for good reason. The native *Viburnum* species of Florida are dependable in their flexibility of shape and size, generous blooming, and shade tolerance. Of the genus, Walter's viburnum (*Viburnum obovatum*) and arrowwood viburnum are most relevant to Florida gardeners.

Arrowwood viburnum (*Viburnum dentatum*) grows wild in wetlands with moderate inundation like swamps, stream banks, and mesic woods. It is deciduous and significantly shorter in stature than Walter's, reaching 12 feet at its highest. Its fruits are a more saturated, metallic hue of blue than Walter's, with some cultivars grown to highlight this differentiating feature. Arrowwood should be chosen for areas with well-draining but moist soils.

Walter's viburnum, the more drought tolerant and common of the two, is most beloved for its versatility in height and shaping. It will reach heights of 20 feet but is typically pruned into much shorter hedges, screens, or shapely accents. It generally enjoys moister soils, being native to the hydric hammocks and floodplain swamps of our state, but will tolerate short periods of drought after establishment. In extreme droughts, the plant may drop leaves but will often regrow them as the rains return.

Both species produce impressive masses of clustered, white blooms, which are a favorite of pollinators. Birds enjoy their concentrated foliage for nesting and cover, as well as the nutritious fruits in the summer and fall. All native viburnums are hosts for the spring azure butterfly (*Celastrina ladon*). Despite attracting so many pollinators, viburnums have relatively little nectar compared to others, so do not use these as your main nectar species in the garden, especially for high numbers of butterflies or bees.

Family: Adoxaceae (Moschatel Family)

Hardiness: Species dependent; *V. dentatum* grows in 8A–9B, *V. obovatum* extends to 10A

Lifespan: Long-lived perennial

Zone: Riparian–Upland

Soil: Occasionally inundated to dry sand, loam, and humus with variability

Depth: Shoreline, no permanent inundation, tolerates longer seasonal flooding

Exposure: Species dependent; *V. dentatum* and *V. obovatum* can tolerate full fun and all prefer partial shade

Growth Habit: Species dependent; *V. dentatum* is generally 6–12 feet tall and *V. obovatum* is around 10–15 feet tall, although it can get much larger under certain conditions

Propagation: Seed, division

The flower of *Itea virginica* is quite literally a "sweet spire."

VIRGINIA SWEETSPIRE

(Itea virginica)

With a compact size and graceful growth habit, the Virginia sweetspire (also commonly referred to as Virginia willow) is a favorite of Florida horticulturists who seek a moisture-loving shrub that thrives in shadier locations. Beginning in early summer, its white, honey-scented flowers dangle from its branches with an elegance that attracts pollinators and humans alike. These blossoms are long lasting, providing nectar sources for pollinating insects after the flush of spring comes to an end. Its dense foliage provides shelter for an assortment of small mammals and birds, and its deep root network prevents erosion. While it grows naturally in low-lying woods and wetland margins, it tolerates a fairly wide range of soil conditions and can do well in sunnier locations.

The shrub tends to produce suckers, so it is better to plant in sites where its spread will not be an issue. Its appearance is a common sight alongside backwater streams and creeks in mixed understory communities. Along with larger shade trees and other stream/shoreline-loving species like leather fern (*Achrosticum danaeifolium*), lizard's tail (*Saururus cernuus*), and arrow arum (*Peltandra virginica*), sweetspire can help add magic, dimension, and erosion control to flowing areas. Its deep shade tolerance and thin, tall stature separate it from many bushier shade species that are commonly chosen for similar areas.

Family: Iteaceae (Sweetspire Family)
Hardiness: 8A–10B
Lifespan: Long-lived perennial
Zone: Riparian–Upland
Soil: Wet—not incredibly dry clay, loamy, or sandy soil

Depth: Shoreline, permanent inundation not deeper than a few inches alongside shallow streams, tolerates longer seasonal flooding
Exposure: Partial shade to full shade
Growth Habit: 8 feet tall and 6 feet wide
Propagation: Seed, cuttings

3
WILDFLOWERS

The flowers of the fireflag are a prime example of why upland wildflowers shouldn't have all the fun! Aquatic wildflowers have just as diverse and uniquely attractive an array of blooms as dry-loving species. A full view of this plant can be found on page 9.

ALLIGATOR FLAG
(Thalia geniculata)

Alligator flag, or fireflag, is a multitalented resident of Florida's ponds, marshes, and swamps. In natural sites where ample conditions have allowed it to develop its full height of 9 feet or more, naturalists can feel as if they've been sent back to prehistoric times when macroflora reigned. The plant forms grand, thick colonies of massive,

As a skipper, the Brazilian skipper is known for its typically quick and jumpy flight pattern, giving it the impression of "skipping" through the sky.

verdant foliage that provide excellent cover for nesting birds and other creatures. Towering even higher above these lance-shaped leaves are the spectacles of flowers that extend on stalks more than a foot long. These flowers range in color from purple to white and are broadly lanceolate. A favorite of the authors, they dangle in such a way that the final inflorescence resembles a display of fireworks frozen in time, zigzagging toward the ground. Some fireflags exhibit red highlighting, particularly where the leaf and stem meet, with several showy cultivars seen worldwide in botanical gardens.

While planting should be restricted to areas where viewing isn't an issue, alligator flag excels as a shoreline plant along freshwater bodies where it will provide habitat and food for wildlife and its growth habit will provide erosion control and water filtration benefits, in addition to blocking the growth of invasive species. The alligator flag is a larval host for the Brazilian skipper butterfly (*Calpodes ethlius*). Interestingly, the caterpillars will emerge from the eggs and make shelters from the leaves, folding and securing them with silk. At night, these youngsters will emerge to feed (often causing moderate defoliation in the process, although this is usually not an issue).

Family: Marantaceae (Arrowroot Family)

Hardiness: Zones 8A–10B

Lifespan: Long-lived perennial

Zone: Littoral–Riparian

Soil: Aquatic to consistently wet loam, muck, and sand

Depth: 18 inches to wet soils; it handles seasonal flooding well and large established groups can handle even deeper water in the rainy season

Exposure: Full sun to partial shade

Growth Habit: 6–9 feet tall and 3–6 feet wide

Propagation: Seed, rhizomes

The delicate blossoms of the duck potato, in tandem with the vibrant stalks of the pickerelweed and handsome stands of spikerush, create a visually textured and aesthetically engaging living shoreline garden.

ARROWHEADS

(Sagittaria spp.)

One of the "big three" emergent natives to be planted along shorelines, arrowhead (or duck potato) refers to both *Sagittaria latifolia* and *Sagittaria lancifolia*. With its dainty ivory blooms and dense, tuberous root network perfect for protecting banks from erosion, arrowhead should be a staple of pond and lake shorelines across the state. Growing wild in a variety of wetland habitats, arrowhead is a beloved jewel of the waters for birds, pollinators, and aquatic creatures. Despite its name, ducks don't actually eat its corms (they're much too large and deeply buried), but they do enjoy its seeds and fruits, alongside other birds and wildlife, which take cover among its broad, arrow-shaped leaves. *Sagittaria latifolia* can be distinguished from *Sagittaria lancifolia* by taking note of its more broadly ovate leaf structure, while the latter has a much more lance-like form with the leaf connecting seamlessly to the stem—hence its botanical name. *Sagittaria lancifolia* is a mainstay alongside pickerelweed and knotted spikerush as a most frequently used plant in shoreline restoration. In the authors' experience, this species may be resistant to some herbicides, such as glyphosate, though this result could be due simply to its thick leaf cuticles. Although many larger specimens have corms that are edible to humans and have been used as a food source historically, be aware that plants grown in residential areas or wild locations near roads uptake pollutants and other compounds from runoff and applied chemicals.

A less popular *Sagittaria* species is the grassy arrowhead (*Sagittaria graminea*). Its common name refers to its slender, grass-like leaves that can occasionally take on the arrowed shape of its relatives, although this is not typical. It shares the characteristic three-petaled white flower with a contrasting yellow center of the other listed species. *Sagittaria subulata*, or dwarf duck potato, is a rather minute species that looks like underwater grass with its tiny, ovoid leaves. It does well in muddy areas or locations with water fluctuations and can tolerate up to several feet of clear water.

Family: Alismataceae (Water Plantain Family)
Hardiness: Zones 8A–10B
Lifespan: Long-lived perennial
Zone: Littoral–Emergent
Soil: Aquatic to consistently wet muck and pond bottom

Depth: 3 feet is the maximum depth *S. lancifolia* can grow in long term, but most *Sagittaria* do best in 6 inches to 1 foot of water and can handle short dry periods in which they usually drop their seeds
Exposure: Full sun
Growth Habit: 3–4 feet tall

The authors celebrated finding this mammoth climbing aster growing wild roadside. It is truly outstanding what kind of sprawl is possible for this plant!

ASTERS

(Symphyotrichum spp.)

There is a joke among plant biologists when one is having trouble identifying a species; another one will say matter-of-factly, "Ah, an aster." This comment is tongue-in-cheek, as there are over 440 asters in Florida, 41 of which are endemic to the state. There are a handful that carry the "aster" designation in their common title and which are of particular interest to the Florida gardener.

Climbing aster (*Symphyotrichum carolinianum*) is a rapidly growing, sprawling "shrub" notable for its opulent pinkish purple flowers. It can be pruned to any shape but is arguably most attractive when allowed to sprawl along the slopes of ponds and other wetland habitats, drawing a plethora of pollinators in with its fall- and winter-flowering blooms. On mature plants, this blossoming period is remarkable, painting a lavender flourish across the landscape.

Unlike most species in its genus, the climbing aster is a vining shrub-like plant with large tendrils growing more than 10 feet long. It is capable of growing in unusual areas and is often seen along riverbanks, in dying palm hearts, among cabbage palm boots, or in the riprap along lakes! Unlike many woody plants, it is not particularly apt at controlling erosion and should be used instead to attract pollinators and add color to steep or sheer surfaces where its hanging nature can be on full display. It is a host for the pearl crescent butterfly (*Phyciodes tharos*).

The pearl crescent butterfly is one of the most abundant butterfly species on the east coast of the country, with a speedy and low-lying flight pattern that allows easy spotting if you've got a quick eye.

Like the climbing aster, the bushy aster, or rice button aster (*Symphyotrichum dumosum*), is a profuse bloomer of nearly identical flowers, but it has a significantly more contained growth habit of only around 3 feet tall. Elliott's aster (*Symphyotrichum elliottii*) shares the flowers and abundant nature of its listed relatives. In the fall, its fragrant and concentrated blossoms dot roadsides and swamps alike, providing much-needed seasonal food for pollinators. Elliott's aster stands apart for its height (typically around 4 feet) and tendency to form stands.

Family: Asteraceae (Sunflower Family)

Hardiness: Listed species grow from 8A–11, except for *S. carolinianum*, which grows as far south as 10B

Lifespan: Long-lived perennial

Zone: Species dependent, but generally riparian–upland

Soil: All species listed tolerate sand and more broadly clay and loam, but moisture will vary

Depth: *S. carolinianum* and *S. elliottii* are tolerant of inundation and light seasonal flooding, though they generally prefer to enjoy consistently moist soil; *S. dumosum* prefers slightly drier conditions and can tolerate extreme drought, while also enjoying regularly moist soils

Exposure: Full sun to partial shade

Growth Habit: Species dependent

Propagation: Seed

Bidens alba might just be the most underappreciated native of Florida.

BIDENS

(Bidens spp.)

Of the *Bidens* species native to Florida, *Bidens alba* is by far the most well known, although not for the best reasons. *Bidens alba*, or beggarticks, is recognized as a weed, with many native plant experts then raising the question "Well, what exactly *is* a weed?" *Bidens alba* is certainly a reproductive expert, seeding rapidly and fruitfully throughout the ecosystem where it is found. The seeds have barbs that attach efficiently (albeit annoyingly) to clothing and fur. If a weed is generally considered a plant incompatible with human desires and environmental goals, beggarticks can be considered as such for its refusal to stay agreeably within the bounds of its original placement. Perhaps the Florida gardener can reconsider this weedy status in light of the plant's competitive ecological benefits. It is a favorite of pollinators and a larval host for the dainty sulphur butterfly (*Nathalis iole*), in addition to being the third most common source of nectar for honey production in Florida. While its prolific nature may be unattractive for certain humans, the creatures who rely on it most definitely appreciate its continued abundance. Mowing and pulling are two methods to keep *Bidens alba* at bay for those who want to moderate its stretch.

    Bidens alba isn't the only member of the genus that deserves a mention. Burr marigold (*Bidens laevis*) is a densely growing wildflower found in wetland ecosystems that boasts bright yellow flowers throughout the seasons, with longer growing seasons as you move farther south. Its growing habit provides a shrub-like appearance, with individuals growing as tall as 6 feet depending on site conditions. However, they tend to bend the taller they reach, so planting alongside grasses that provide support for the blossoms is a good idea in the landscape setting. *Bidens mitis*, or smallfruit beggarticks, is a visually similar relative whose blossoms shroud the plant in a flaxen haze and is also the host plant for the cloudless sulphur butterfly (*Phoebis sennae*). Both wildflowers enjoy full to partial sun. *B. laevis* is best for aquatic gardening, as it has a high flood tolerance and can grow as far south as zone 11, while *B. mitis* will be better in areas with more prolonged dry periods comparatively. Do not worry about planting *B. alba*; it will find you!

Family: Asteraceae (Sunflower Family)

Hardiness: *B. laevis* grows in zones 8A–10B, with *B. mitis* extending only to 10A

Lifespan: Short-lived perennials, sometimes annuals in cooler areas

Zone: Riparian–Upland

Soil: Species dependent, though *B. laevis* and *B. mitis* both enjoy extremely wet muck or sand

Depth: Species dependent, though typically shoreline, no permanent inundation, tolerating longer seasonal flooding up to several inches

Exposure: Generally full sun to partial shade

Growth Habit: 1–3 feet tall

Propagation: Seed

The vibrant stalks of blazing stars bring a reviving spiritedness to the fall garden.

BLAZING STARS

(Liatris spp.)

There are sixteen species of blazing star native to Florida, something we should all be grateful for. These natives live up to their dreamlike name, producing lavender stalks from basal rosettes. They are a favorite in bouquets for these special inflorescences, which are as desired by humans as they are by pollinators. These plants require very little space but arguably look best when planted in clusters. Four of the sixteen species are commonly cultivated: Chapman's blazing star (*Liatris chapmanii*), dense blazing star (*Liatris spicata*), evergreen blazing star (*Liatris laevigata*), and graceful blazing star (*Liatris gracilis*). Most have unremarkable foliage until flowering, although the evergreen blazing star maintains its pronounced basal rosette through the winter.

Dense blazing star is best for aquatic gardens, growing wild in mesic to wet flatwoods, bogs, roadside ditches, and similar wetland habitats. It requires little maintenance once established, although it will enjoy structural support from grass species or similarly structured wildflowers like goldenrods. The remaining blazing stars are more drought tolerant, with Chapman's enjoying very well-draining, dry soil. Evergreen and graceful blazing stars are found wild in dry to mesic flatwoods and have broader tolerances of moisture conditions. Like many tall and thin wildflowers, they benefit from having support from nearby grasses or shrubs to help support their top-heavy weight. Wiregrass (*Aristida stricta*), bluestems (*Andropogon* spp.), beaksedges (*Rhynchospora* spp.), carex sedges (*Carex* spp.), and any other species suitable to the area with long, thin blades or stems would be great choices for such a purpose.

Family: Asteraceae (Sunflower Family)
Hardiness: Zones 8A–10B, depending on species
Lifespan: Long-lived perennial
Zone: Upland
Soil: Species dependent

Depth: Shoreline only, no prolonged inundation
Exposure: Full sun
Growth Habit: Species dependent, but typically less than 3 feet tall
Propagation: Seed, division

Arguably the most recognized skipper in North America, the silver-spotted skipper displays "nectar robbing" behavior, a phenomenon in which a species consumes the nectar from a plant without pollinating it. While this practice may sound extreme, the actual impact of this behavior on pollination and overall plant health varies drastically.

The alternative common names of bonesets, including fever-wort and sweating plant, refer to the plant's historical medicinal use as a diaphoretic.

BONESETS

(Eupatorium spp.)

Bonesets resemble a white version of blue mistflower in terms of appearances and growth habit but are limited in location to northern counties. Indigenous Americans used the plant to treat fevers and common colds and introduced the species to colonists for this reason. These winter-dormant wildflowers can sometimes be considered weedy when not in bloom due to their height, but when planted alongside other native species, this effect is negligible. The triangular leaves, which are venated and covered in tiny hairs, seem to wrap around the stalk.

Birds and other wildlife consume the fruit of the plant, and it is especially important for native bees. Common boneset (*Eupatorium perfoliatum*) is found growing wild in many moist places where sunlight is dappled, such as along the edges of streams or rivers. Some, like *E. mohrii* or *E. rotundifolium*, are more adapted to water, while others are more common in scrublands. These species do not transplant well from full-sized plants but do wonderfully in seed mixes, as well as self-seeding regularly. Like all species in the aster family, they require open soil to reseed on, although they can flower in a variety of circumstances and tolerate occasional inundation.

Dog fennel (*Eupatorium capillifolium*) is a common species that, while native, is often seen as a pest with its aggressive proliferation, noxiousness to livestock, and debatably low pollinator benefit. Sources differ on the latter, but it is certain that other plant species are far more apt at serving our native insects. Mowing or cutting prior to seeding will help reduce the thick stands they often form. Planting is not recommended, but keeping small groupings of the species where they naturally emerge can be useful for privacy and maintaining biodiversity.

Family: Asteraceae (Sunflower Family)

Hardiness: *E. perfoliatum* is limited to 8B–9A; other bonesets are found throughout Florida but less commonly cultivated

Lifespan: Short-lived perennial

Zone: Riparian–Upland for *E. perfoliatum*, *E. mohrii*, and *E. rotundifolium*; other bonesets tend to be more suitable for strictly upland habitats

Soil: *E. perfoliatum* enjoys wet to somewhat moist clay, loam, and sand

Depth: Shoreline, no permanent inundation, tolerates longer seasonal flooding

Exposure: Full sun to partial shade for many bonesets, although this is species dependent

Growth Habit: Species dependent

Propagation: Seed, division

With a host of ecological benefits and human uses, one will want to savor Browne's savory wherever it is found.

BROWNE'S SAVORY

(Clinopodium brownei)

Browne's savory earned its name—this evergreen groundcover smells and tastes strongly of mint. Similarly to water hyssop, the native enjoys forming mats of green foliage in moist areas like wetlands, ditches, and shorelines. The plant touts miniscule but structurally interesting lavender flowers with vibrant purple "throats" that almost seem to serve as landing strips for small pollinating insects. Many flowers dip into the UV spectrum for their flower colors to attract pollinators, sometimes not being visible to humans. Still others may mimic the look of certain pollinating insects, show off bright contrasts to lure them, or even further imitate other, more preferred flowers.

Like bacopas, Browne's savory enjoys sandier soils and full sun, and it is able to withstand long-term inundation. They are both popular aquarium plants, but their true beauty, like many wetland species, relies on water *changes*. These species reproduce vegetatively in deeper water, but only when exposed on drying banks will they flower and seed. Use these species on exposed banks to help reduce erosion whole covering open ground with swaths of delightful greenery. As an added bonus, many small fish, invertebrates, and amphibians rely on these plants to shelter their eggs and young.

Family: Lamiaceae (Mint Family)

Hardiness: Zones 8B–10B

Lifespan: Long-lived perennial

Zone: Riparian–Upland

Soil: Consistently wet to somewhat moist loam and sand

Depth: Shoreline to an aquatic form but prefers 2 inches or less of water that dries out periodically

Exposure: Full sun to partial shade

Growth Habit: 6 inches tall and several feet wide

Propagation: Division, cuttings

Like many native species with "button" in the name, the button snakeroot has globular blossoms that are quite popular with pollinators.

BUTTON SNAKEROOT
(Eryngium yuccifolium)

Sometimes referred to as rattlesnake master, the button snakeroot has as coarse an appearance as its name may suggest. Its succulent leaves are marginally covered in spines, and its blossoms are clusters of prickly white globes that shoot formidably from erect stems. For this reason, the plant is not very popular with mammalian herbivores. However, the button snakeroot has the advantage of attracting many kinds of insects, particularly those that prey on garden pests, serving as a natural line of defense against those that might wish to inflict harm on your landscape.

The button snakeroot is native to all counties in Florida, growing wild in wet and coastal flatwoods, wet prairies, and savannas. However, if the soil stays too wet for too long, it is threatened by root rot. For a similar but more site-specific species, explore *Eryngium aquaticum*, the uncommon corn snakeroot that displays stunningly blue globes.

Recent research suggests the *Eryngium* genus attracts some of the greatest diversity of pollinators in the state. While this book focuses on butterflies, planting button snakeroot and similar species can ensure that you see more beetles, bees, wasps, and other insect pollinators in your community.

Family: Apiaceae (Carrot, Celery, or Parsley Family)

Hardiness: Zones 8A–11

Lifespan: Long-lived perennial

Zone: Riparian–Upland

Soil: Consistently wet to dry muck and sand

Depth: Shoreline, no permanent inundation, tolerates longer seasonal flooding

Exposure: Full sun to partial shade

Growth Habit: 3–5 feet tall

Propagation: Seed, division

Fiery skippers, like many other skippers, often hold their wings in a triangular shape that is thought to be best for absorbing the sun's rays or helping them move through the grasses on which their larvae feed. Many skippers are the "pests" of lawn grasses such as Bermuda, so remember—when you spray, the butterflies go away!

The resemblance of cardinal flowers to the vestments of Roman Catholic cardinals is more than apparent from a perspective like this one.

CARDINAL FLOWER

(Lobelia cardinalis)

Threatened in the state of Florida, the cardinal flower is a statuesque, boldly scarlet herbaceous wildflower. Its tubular red flowers are structurally ideal for hummingbirds and similarly long-throated butterflies such as swallowtails and other pollinating insects. It is said that the common name is derived from the blooms' resemblance to the robes of the Roman Catholic cardinals. The long stalks of flowers will poke out during the fall, while most of the year it exists as an unassuming, short herb with dark purplish green leaves.

Growing wild in swamps, bogs, and similar wetland habitats in central and northern counties, cardinal flower is a striking and beneficial addition to the aquatic butterfly garden, either as an accent or in clumps. Unlike most *Lobelia* species, it grows just fine underwater for very long periods and is occasionally even used as an aquarium plant. To flower and reseed, they do require dry land, where they will easily spread their millions of tiny, dust-like seeds. Cardinal flower generally does best in moist, partial shade environments, particularly with moving water and well-draining soils that resemble the slow-moving riverbanks and shallow seeps to which they are native.

Stagnant water reduces growth, even if growing underwater, and they are not sufficiently fast growing to outgrow algae like many other aquatic species. Be mindful that many local ecotypes may be more or less water tolerant, so, if needed, slowly adapt your species to wetter or dryer ecosystems and always try to start in locations where your specimen will keep moist to wet feet without submersion.

Family: Campanulaceae (Bellflower Family)
Hardiness: Zones 8A–9B
Lifespan: Short-lived perennial
Zone: Riparian–Upland
Soil: Aquatic to somewhat moist loam, muck, sand, or pond bottom

Depth: Several feet underwater in clear, sunny areas but needs dry land to flower
Exposure: Full sun to full shade
Growth Habit: 3–5 feet tall when flowering, usually short herbs under 6 inches when not
Propagation: Seed

The flowers of the Carolina redroot look rather fleecy in appearance, with whitish outer petals encasing much sunnier yellow petals.

CAROLINA REDROOT
(Lachnanthes caroliniana)

When not in bloom, redroot looks like a tall grass, forming a relatively dense groundcover along the marshes, flatwood, disturbed sites, and other wetland habitats where it is found wild. In the summer and fall, it reveals its fan-like clusters of cream-colored flowers smothered in wooly hairs. These inflorescences are a favorite of butterflies and other pollinating insects. Its seeds (a special treat for the endangered sandhill crane) are eaten by birds, and its vibrantly red roots are dug up and eaten by hogs, often leading to large amounts of rutting where hogs and redroot overlap.

The plant is fast growing and multiplies via rhizomes, establishing hardily under the right conditions. It is an essential to many wet grasslands or open spaces in moist, wooded areas. Its unusual flowering pattern makes it a fun addition to a mixed rain garden. Land managers often dislike redroot, as it can be difficult to differentiate from cattails and is constantly confused with many *Iris* species. The trick to identifying young cattails, irises, and redroot is simple: redroot has red roots and a flat base, irises have white roots and a flat base, and cattails have white roots and a round base. You will also typically see redroot and cattails in large clumps, whereas irises are more often seen as individuals or in smaller clumps and are more common in shadier areas.

Family: Haemodoraceae (Bloodroot Family)

Hardiness: Zones 8A–10B

Lifespan: Long-lived perennial

Zone: Riparian

Soil: Consistently wet to somewhat moist muck, limerock, and acidic sands

Depth: 6 inches or less of consistent water can be tolerated, but it does best in wet soil

Exposure: Full sun to partial shade

Growth Habit: 3 feet tall

Propagation: Seed, division

In Florida, the gulf fritillary has two well-documented migrations. The first occurs in the spring, with the species flying northward. The second occurs in the fall, with the species flying southward.

The leaves of the coontie are easily identifiable and exceptionally sturdy in the landscape, delivering a prized tropical aesthetic to the understory.

COONTIE

(Zamia integrifolia)

The only cycad native to Florida, coontie is most known for its status as the sole larval host of the endangered atala butterfly (*Eumaeus atala*). In the early 1900s, coontie was overharvested by those using its starchy root to make flour, which was included in everything from World War II ration packs to animal crackers. Atala populations consequently collapsed and remained limited to the southern tip of the state—until recently! Over the years, the coontie has been integrated into home and urban plantings, cherished for its hardiness and glossy, featherlike leaves that resemble a sort of trunkless palm. In the last decade, atala populations have risen significantly and expanded farther north, benefiting both from the reestablished coontie populations and from the ever-warming climate.

Coontie can be planted in full sun or shade and is moderately salt tolerant, drought tolerant, and cold hardy, putting this shrub in a class of its own for more than just its Mesozoic roots. These plants take a decade to reach mature size, so they can be pricier than other natives but are certainly worth every penny. This slow growth rate makes these plants very low maintenance, with their consistent height preventing the need for the trimming usually required by hedges. Their dense, clumping habit also reduces the need to weed.

The endangered atala butterfly is making a newsworthy comeback thanks to the widespread use of coontie in landscaping.

Family: Zamiaceae (Sago-Palm Family)
Hardiness: Zones 8A–11
Lifespan: Long-lived perennial
Zone: Upland
Soil: Briefly inundated to extremely dry sand

Depth: Tolerant of moist, well-drained soils, including saltwater
Exposure: Full sun to full shade
Growth Habit: 2–3 feet tall and 3–5 feet wide
Propagation: Seed

The quirky flowers of the Florida hedgenettle make a lively appearance in a "rewilded" lawn.

FLORIDA HEDGENETTLE

(Stachys floridana)

Prior to the 1940s, it was believed that the Florida hedgenettle (also known as Florida betony) was entirely endemic to our state. To other states' misfortune, the aggressive and hard-to-remove groundcover soon began its spread into the Southeast. For this development, it receives a bad rap, and understandably so. However, for those of us lucky to call it a native, the Florida hedgenettle is a perfect choice for meadows and gardens, as well as a replacement for turf grass. It is recommended only for sites where its exuberant spread is encouraged. In those spaces in which it is welcome, it provides a showy display of tubular, pinkish to purplish white flowers that grow on spikes. Its underground network of stems produces cream-colored, edible tubers that resemble a rattlesnake's rattle and have a mildly sweet taste and crunchy texture (much like its relative, the Chinese artichoke).

Florida hedgenettle can be mowed without harm and is tolerant of all sun conditions. It is found growing wild in moist, disturbed sites and flatwoods throughout much of Florida. Since it is so aggressive and possesses particularly strong rhizomes, it is one of the few species that can compete successfully with invasive shoreline plants. Instead of planting a "no mow zone," consider a mass planting of Florida hedgenettle and similar shorter stature natives.

Family: Lamiaceae (Mint Family)
Hardiness: Zones 8A–10B
Lifespan: Long-lived perennial
Zone: Riparian–Upland
Soil: Occasionally inundated to not extremely dry loam and sand

Depth: 6 inches or less
Exposure: Full sun to full shade
Growth Habit: 1 foot tall
Propagation: Tuber

Frogfruit is one of the most standout alternatives to turfgrass for the Florida lawn. As a plus, it's a larval host for a handful of butterflies, which will certainly appreciate its integration into your landscape.

FROGFRUIT
(Phyla nodiflora)

Tired of spending thousands of dollars annually fertilizing and mowing a lawn of non-native turf grass? Consider an alternative: the hardy, evergreen frogfruit grows no more than 6 inches and is the larval host for the common buckeye (*Junonia coenia*), Phaon crescent (*Phyciodes phaon*), and white peacock (*Anartia jatrophae*) butterflies. Not only is it a phenomenally contributive addition to the butterfly garden, but its dense, sprawling growth habit also stabilizes sediment. For a no mow Florida yard, it is ideal to mix with sunshine mimosa (*Mimosa strigillosa*) in drier locations or water hyssop (*Bacopa caroliniana*) in wetter or poorer draining areas. It does well on its own in hanging pots, as it takes up a whimsical vining nature when grown off the ground, with its purple and white matchstick-like flowers blooming reliably year-round.

Also called turkey tangle frogfruit, this versatile perennial is found wild throughout the entirety of our state in hammocks and along roadsides. However, it is extremely resilient under many conditions and finds its way naturally to a variety of habitats, from dry scrublands to underwater locations.

Family: Verbenaceae (Verbena Family)
Hardiness: Zones 8A–11
Lifespan: Long-lived perennial
Zone: Riparian–Upland
Soil: Occasionally inundated to not very dry clay, loam, and sand

Depth: Shoreline, no permanent inundation, tolerates longer seasonal flooding
Exposure: Full sun to partial shade
Growth Habit: 6 inches tall
Propagation: Cuttings

In Greek mythology, Phaon was an ugly, ancient boatman. After ferrying Aphrodite a great distance and refusing payment in return, she gifted him beauty and youth. While the Phaon crescent spans less than an inch long, its bold, intricate markings make it plain to see how it earned such a name.

Frostweed has more experience with ice sculptures than most native Floridians themselves!

FROSTWEED

(Verbesina virginica)

You might be thinking frostweed is a peculiar name for a plant native to Florida, and you'd be right. Its curious designation is well earned, though. In the less tropical winters of northern counties, the herbaceous perennial has a habit of exuding water from its stems. When this water freezes, the result is a bizarre display of ribbon-like "ice sculptures." Not too bad for a Florida native!

Besides its wintery wonders, frostweed is known for its tall display of white flowers with noticeably contrasting purplish black anthers. These blooms emerge in the late summer and last through much of the fall. In the wild, frostweed grows on the margins of moist forests and hammocks. The plant would do best planted *en masse* in the landscape in a location where its rather tall and occasionally weedy appearance is welcome.

This taller, shade-loving species provides lovely splashes of white and does well in a mixed planting of beautyberry (*Callicarpa americana*), shiny wild coffee (*Psychotria nervosa*), and marlberry (*Ardisia escallonioides*). Its flowers are less impressive than many asters, but the leafy stems provide a unique identifying characteristic.

Family: Asteraceae (Sunflower Family)
Hardiness: Zones 8A–10B
Lifespan: Long-lived perennial
Zone: Upland
Soil: Occasionally inundated to not extremely dry clay, loam, and sand

Depth: Shoreline, no permanent inundation
Exposure: Partial shade to full shade
Growth Habit: 4–5 feet tall
Propagation: Seed, division

Luckily for the Florida gardener, ironweed takes its name from its hardiness. Its stem is tough, and it often reseeds heavily—a welcome establishment for those interested in beautiful flowers and the pollinators that come along with them.

GIANT IRONWEED

(Vernonia gigantea)

The rightfully named giant ironweed takes up residence in the wet to mesic pine flatwoods, stream banks, and forest margins of Florida. Flowering in the summer and fall, the robust wildflower extends bunches of delightfully bright purple blooms among its narrow, ovate green leaves. It is popular for its effectiveness as an aggressively growing and adaptable wetland flower. While it will not tolerate drought and dies back in the winter, it is still rather hardy and adventitious, producing suckers that would be welcome in a naturalistic landscape.

Giant ironweed is the larval host to the spring azure butterfly (*Celastrina ladon*) and attracts a wide variety of other pollinating insects. Many smaller species of ironweed are also found in Florida, with Florida ironweed (*Vernonia blodgetti*) and narrowleaf ironweed (*Vernonia angustifolia*) being the most common. Ironweeds do poorly in pots and are best grown from seed, making themselves a staple of moist meadow wildflower mixes. Like many plants in the sunflower family, they are staple generalists for pollinator communities.

Family: Asteraceae (Sunflower Family)

Hardiness: Zones 8A–10B

Lifespan: Long-lived perennial

Zone: Upland

Soil: Occasionally inundated to not extremely dry loam and sand

Depth: Shoreline, no permanent inundation

Exposure: Full sun to partial shade

Growth Habit: 4–6 feet tall

Propagation: Seed, division

The swarthy skipper doesn't have a ton going for it in comparison with other butterflies. It is quite small and not too common, and it has no distinctive markings. Perhaps, though, one can find beauty in its humble, earthy simplicity.

Despite its softly golden, stylish flower, the golden canna lifts a heavy load when it comes to erosion control, nutrient balancing, and pollinator benefit.

GOLDEN CANNA
(Canna flaccida)

Golden canna is one of the "big three" aquatic species for building living shorelines along ponds and lakes. Its powers of erosion control and shoreline stabilization are hard to beat thanks to its intricate and dense rhizomatic network. From verdant lance-leaved foliage sprout robust yellow flowers that add valued color and charm to any landscape. Often, this gorgeous native is referred to as the "bandana of the Everglades." While golden canna does well fully immersed, it can also tolerate simply having "wet feet." The plant contains phytoremediation agents that aid in removing nitrogen and phosphorus from the water, two nutrients that, in excess, can lead to ecosystem imbalances resulting in fish and wildlife kills, harmful algae blooms, and more hazards. This species is the larval host for the Brazilian skipper butterfly (*Calpodes ethlias*). Its dense growth works to block out the emergence of invasive species like torpedo grass. In the authors' experience, planting thick stands of canna is one of the single best ways to replace no mow zones and promote thorough living shorelines with impeccable erosion control.

Golden canna's submerged structures are utilized by fish and invertebrates for cover, while the above-water portions are similarly valued by birds and other creatures. Historically, the stems and leaves have been used as feed for cattle, and the leaves can be processed to make a beige paper. Indigenous Floridians used golden canna seeds to make jewelry, dyes, toys, and ceremonial objects. For growing, however, these seeds are extremely tough to germinate, requiring acid baths and scarification, so stick to bare root division for easy propagation.

Family: Cannaceae (Canna Family)
Hardiness: Zones 8B–11
Lifespan: Long-lived perennial
Zone: Emergent–Riparian
Soil: Aquatic to occasionally inundated clay, loam, muck, and sand

Depth: 18 inches during seasonal flooding but 6 inches to start new bare root plants
Exposure: Full sun to partial shade
Growth Habit: 3–6 feet tall with size very determinate on sun exposure
Propagation: Seed is difficult, division or bare root is easiest

If you forgot the goldenrod's common name, simply describing the plant visually would probably get you close enough!

GOLDENRODS

(Solidago spp.)

Of the twenty-two species of goldenrod native to Florida, around half of them can be found in the nursery trade. The four most commonly cultivated are seaside goldenrod (*Solidago sempervirens*), wand goldenrod (*Solidago stricta*), pinebarren goldenrod (*Solidago fistulosa*), and Chapman's goldenrod (*Solidago odora* var. *chapmanii*). Goldenrods are straightforwardly named, distinctive for their extended inflorescences of golden, tubular blooms. These perennials have basal leaves year-round, but when it's time for flowering, a stem ascends from this hearty base to produce the nodding wands. The plant is mistakenly blamed for the fall allergy season, of which ragweed (*Ambrosia* spp.) is the true culprit. The goldenrods are attractive to pollinators, especially monarch butterflies, fireflies, and the goldenrod beetle (*Chauliognathus pensylvanicus*), which specializes in pollinating goldenrods.

Common buckeye butterflies are rather solitary. When multiple caterpillars feed on the same plant, this process is not done cooperatively or as part of a grouped behavior, as is common with many species. They are easily identified thanks to their bold eyespots that may serve to startle predators.

Seaside goldenrod grows wild along coastal sites like dunes and brackish marshes. The plant grows 4–6 feet tall and spreads via rhizomes to form colonies. It is drought tolerant and does best in full sun conditions. Wand goldenrod is a more moisture-loving species, found natively in mesic flatwoods and prairies throughout the state. It is shorter than the seaside goldenrod, reaching only around 2–4 feet in height. Pine-barren goldenrod is similarly short in stature, growing up to 3–5 feet. It is the most common native goldenrod, found everywhere from wetland habitats to dry, upland flatwoods. Chapman's goldenrod is around the same height and enjoys sandhills and dry hammocks.

Family: Asteraceae (Sunflower Family)
Hardiness: Varies with species, but found throughout the state
Lifespan: Species dependent, typically perennials
Zone: Upland
Soil: Typically sandy, with some species tolerating loamier conditions

Depth: Shoreline, no inundation, though sometimes 2 inches or less of seasonal rain
Exposure: Full sun to partial shade
Growth Habit: Variable in size from 2 to 12 feet, but always as a single herbaceous stalk
Propagation: Seed, division

The seaside heliotrope's rather rare foliage hue and unique structure makes it a treasure of any garden.

HELIOTROPES
(Heliotropium spp.*)*

Seaside heliotrope's foliage, curved flower clusters, and growth habit make it a rare treasure within any landscape it is a part of. Like many species native to the moist, salty areas of our coastline, the evergreen herb has a succulent-like, fleshy leaf. This foliage is unusual in its color, which is somewhere between silver and bluish green. The hue is eye catching and makes the plant easily identifiable for those who have encountered it once before. To top it off, its double-rowed white flowers emerge along a deeply curved spike, earning it the name "monkey tail" in some areas throughout the world.

The seaside heliotrope is a prime nectaring plant for the extremely rare Miami blue butterfly (*Clyclargus thomasi*) as well as many of the ceraunus butterflies, like the ceraunus blue (*Hemiargus ceraunus*). It attracts many other butterflies and pollinating insects as well. For the gardener looking for a more inland species, look to the scorpion's tail (*Heliotropium angiospermum*). It thrives in similar environments but tends to be more aggressive in its growth and can handle loamier soils. For a rare treat that is just beginning to enter the native plant trade, consider sea lavender (*Heliotropium gnaphalodes*), a beautiful and large shrub with pure silver foliage and the largest *Heliotrope* flower clusters native to the state. Unfortunately, this species is endangered and entry into cultivation is slow and expensive, but it is more than worth the wait. All *Heliotropes* listed here prefer sandy, well-draining soil and do best in full sun locations.

Family: Boraginaceae (Borage Family)
Hardiness: Zones 8A–10A
Lifespan: Annual
Zone: Upland
Soil: Occasionally inundated to not extremely dry sand

Depth: Shoreline
Exposure: Full sun to partial shade
Growth Habit: 1 foot tall and 4 feet wide
Propagation: Seed or cuttings

Largely due to urban development and the destruction and fragmentation of habitat, the Miami blue is one of Florida's most endangered insects.

There are few plants that evoke a sense of tropicality as much as the hibiscus.

HIBISCUSES

(Hibiscus spp.*)*

The grandiose flowers of the hibiscus are perhaps one of the most celebrated tropical features of Florida's lush environment. While many nonnatives rule home landscapes, it is the natives that adorn the wild wetlands of our state and contribute the most positively to our ecosystems. Hibiscus flowers vary greatly in color, but all are highly desired by butterflies and hummingbirds. Estimates vary on just how many butterflies and moths utilize them, but many claim at least twenty-six or more species use hibiscuses as a larval host, with some standouts including painted ladies, checkered skippers, the mallow scrub hairstreak (*Strymon istapa*), and the gray hairstreak (*Strymon melinus*). The rose mallow bee (*Ptilothrix bombiformis*) is also a specialist pollinator to this group.

The scarlet hibiscus (*Hibiscus coccineus*) is one of the state's largest and showiest wildflowers. Its gigantic, fiery crimson blooms (see page 5 for a full view) stay open for only a day, but no worries! The plant will produce many flowers throughout the summer. While the species is found wild in sites with moist or inundated soils, it can tolerate well-drained sites given enough moisture. It can grow up to 8 feet tall and will die back in the winter due to dry or cold weather.

The pineland or comfortroot hibiscus (*Hibiscus aculeatus*) is a more mildly drought-tolerant, longer-blooming alternative that sports creamy white flowers atop a shrubbier form that grows to heights of around 3 feet. It flowers typically from the late spring through the fall and will usually self-seed.

Tropical checkered skippers often fly alongside their siblings, the common and white checkered skippers. These tiny butterflies enjoy nectaring from small flowers that are often white in color.

Swamp rosemallow (*Hibiscus grandiflorus*), another wetland-centric native hibiscus, is commonly found growing in colonies along spacious marshes. It blooms from summer into fall, revealing blushing pink flowers. It is quite large, reaching 10 feet tall or more under the right conditions.

Family: Malvaceae (Mallow Family)
Hardiness: Varies heavily by species: *H. coccineus* 8A–10B, *H. grandiflorus* 9A–11, and *H. aculeatus* 8A–9B
Lifespan: Long-lived perennial
Zone: Riparian
Soil: Varies, but usually loam, or sand, with *H. coccineus* enjoying muck and floating wetlands

Depth: Shoreline, no permanent inundation, tolerates longer seasonal flooding
Exposure: Full sun, though some species or individual specimens will tolerate partial shade
Growth Habit: Species dependent
Propagation: Seed

It might be called lizard's tail, but the racemes of *Saururus cernuus* are certainly larger than the tails of the tiny anoles that dart through our landscapes—and we're grateful for that!

LIZARD'S TAIL

(Saururus cernuus)

In the shallow, shaded swamps, marshes, and wet forests of Florida lie colonies of an aquatic wildflower that, while silent, calls out with its novel, distinctive blooms. The lizard's tail produces loads of tiny white flowers organized on elongated, gracefully drooping racemes that are relatively large for its size. The "nodding" nature of these inflorescences calls to mind the curve of a lizard's tail—hence its name. With its shade tolerance, nativity to sites with persistent moisture and occasional flooding, and tendency to form thick patches, this species is ideal for landscapes with canopy coverage and room for the plant to work its restorative magic. Of the native shade-loving species, lizard's tail has some of the largest and most attractive flowers, reliably blooming in the spring through the fall.

In addition to its use as a nectaring plant, lizard's tail is said to be a favorite perch of dragonflies, and the plant provides food and habitat for foraging birds and other creatures. Growing an understory garden in wet environments tends to be quite difficult, as waters tend to be deeper during times of seasonal flooding, sun is rare, and roots dominate. By using lizard's tail, ferns, cardinal flower (*Lobelia cardinalis*), and white swamp milkweed (*Asclepias perennis*), you can add color and textured diversity to even the shadiest and wettest of swamps.

Family: Saururaceae (Lizard's Tail Family)
Hardiness: Zones 8A–10B
Lifespan: Long-lived perennial
Zone: Riparian
Soil: Consistently wet to occasionally inundated clay, loam, muck, and sand

Depth: 6 inches, preferring consistent soil moisture
Exposure: Full shade or dappled partial shade, though they are prone to burning
Growth Habit: 2–3 feet tall
Propagation: Seed, division

The summer-blooming flowers of the loosestrife are a delight to behold and, while a relatively uncommon find in the nursery, make a fabulous addition to a native garden's collection.

LOOSESTRIFES

(Lythrum spp.)

There are two main native species of loosestrife found in Florida: the winged loosestrife (*Lythrum alatum*), which is a taller, flowering herb, and the Florida loosestrife (*Lythrum flagellare*), a sprawling groundcover.

Winged loosestrife is the commercially available of the two, known for its superior ability to attract a myriad of pollinators, from honeybees to skipper butterflies. In color and growth form, it evokes lavender or rosemary with its multiple extended "winged" stems and slender racemes of star-shaped light purple flowers. Occurring naturally in freshwater marshes, prairies, and wet flatwoods throughout most of Florida, the loosestrife provides summer blooms that are particularly attractive to long-tongued pollinators.

Florida loosestrife, however, is an endangered species endemic to the state. While it is unlikely that the species will ever take off in the trade, it should be preserved whenever possible. It is an iconic feature of the Deep Hole in Myakka River State Park, forming the primary groundcover across much of the floodplain and the Deep Hole itself.

Take care not to confuse the purple loosestrife (*Lythrum salicaria*) with the native species. This variety is a common invasive throughout the entirety of the United States and is threatening Florida ecosystems due to its tendency to form dense monocultures that change hydrology, outcompete native plants, and even alter water chemistry.

Like many species of butterfly, the male Delaware skipper will perch on low-lying vegetation, eagerly awaiting the arrival of a female.

Family: Lythraeceae (Loosestrife Family)
Hardiness: Zones 8A–10B
Lifespan: Long-lived perennial
Zone: Riparian–Upland
Soil: Consistently wet to somewhat moist loam and sand

Depth: Winged loosestrife prefers wet feet and up to 6 inches of seasonal flooding; Florida loosestrife is seasonally flood tolerant to several feet in depth
Exposure: Full sun
Growth Habit: 2–3 feet tall for *L. alatum*; 6-inch groundcover for *L. flagellare*
Propagation: Seed

The mangrove spiderlily is one of the most exquisite of the aquatic wildflowers.

MANGROVE SPIDERLILY

(Hymenocallis latifolia)

It is understandable why the mangrove spiderlily is the most cultivated of the fourteen native *Hymenocallis* species. Dotted along the mangrove swamps, coastal swales, and other wetlands of our state lies the regal, perfumed flower, which blooms from spring through autumn, much to the appreciation of naturalists and gardeners lucky enough to be its neighbors. These aromatic and showy blossoms can stand quite tall and will multiply, so it's best to plant in a space where their moderate growth is welcome. The flowers are actually so fragrant they've led to the adoption of another common name: the perfumed spiderlily.

Unlike most lilies in the *Hymenocallis* group, mangrove spiderlilies are surprisingly drought tolerant and even capable of growing on colonized sandbars and mangrove islands that offer partial shade. This is the most common species available throughout much of Florida; other relatives, such as the woodland spiderlily (*Hymenocallis occidentalis*) and alligator lily (*Hymenocallis palmeri*), occasionally enter cultivation, though this circumstance is heavily location dependent. Sadly, these plants are often poached for their beautiful, exceptionally delicate flowers, which often break on translocation. As long as the proper species is planted in the appropriate climate and habitat, they are reliable and showy wildflowers with high site-specificity. Be sure to use as local a source as possible, as these flowers are extremely locally adapted and have high rates of speciation, much like the oaks and pawpaws of Florida.

Family: Amaryllidaceae (Amaryllis Family)
Hardiness: Zones 9B–11
Lifespan: Long-lived perennial
Zone: Riparian–Upland
Soil: Occasionally inundated to dry loam and sand

Depth: Shoreline to several inches deep, preferring fluctuating depths such as from tidal flow
Exposure: Full sun to partial shade
Growth Habit: 2–4 feet tall and wide
Propagation: Division, bulbs

The sweetly named meadowbeauty is a merry character of the landscape.

MEADOWBEAUTIES
(Rhexia spp.)

Meadowbeauties, or species from the genus *Rhexia*, are delightful wetland wildflowers that provide a pop of color to aquatic gardens and forest understories. These species are found everywhere from wet prairies to flatwoods, with each having its specific preferences and tolerances. *R. mariana, R. cubensis, R. alifanus, R. nuttallii, R. petiolata,* and *R. virginica* are not known for their foliage, which is relatively inconspicuous, but instead for their dainty blooms, which range in color from pure yellow to richly veined lilac. Most *Rhexia*, however, tend to be varying shades of purple in color, with differences in soil, water, age, and other site-specific conditions lending to flower variation even within the same population.

Meadowbeauties are buzz pollinated, falling into the mere 9 percent of flowers that are pollinated using this fascinating technique. Its blossoms have sealed anthers that feature a small opening along the top or sides, which allows for pollen to emerge when the anther is vibrated at specific frequencies. An audible change in buzzing can be heard when a bee lands on its petals! This pollination method is very likely the reason why most *Rhexia* species are differentiated by their flower buds varying in size, shape, and hairs. Otherwise, most *Rhexia* have very similar appearances.

Family: Melastomataceae
Hardiness: Species dependent
Lifespan: Long-lived perennial
Zone: Generally riparian
Soil: Species dependent, but usually preferring sandier locations

Depth: Shoreline, or wet feet with seasonal inundation of usually 2–4 inches
Exposure: Full sun to partial shade
Growth Habit: Species dependent, but generally under 2 feet
Propagation: Seed

The butterfly milkweed displays vibrant orange flowers, but our native milkweeds come in an assortment of colors, including stark white, pale green, and rich pink.

MILKWEEDS

(Asclepias spp.*)*

When it comes to butterfly gardening, there is no plant as popular or recognizable as milkweed. The milkweed has risen to acclaim over the last decade for being the only host plant of the monarch butterfly (*Danaus plexippus*), which is severely threatened by habitat loss. Florida is home to twenty-one native milkweeds, an assortment that boasts drastically varying colors, sizes, and site tolerances. Whether you're looking for a showy, drought-tolerant species that will do well along an entryway or one that can colonize a shoreline, there is a native milkweed to serve the purpose. While not every species can be found in nurseries, a handful are common in native nurseries throughout the state.

In the aquatic butterfly garden, two milkweeds reign: *Asclepias perennis* and *Asclepias incarnata*. Both are commonly referred to as swamp milkweed and occasionally differentiated by the color of their blooms. *A. perennis* flowers are a milky white, gracefully situated atop its erect stem that reaches around 3 feet in height. It does well in containers, as long as the soil is kept moist. *A. incarnata* can grow twice as tall, flaunting blooms that vary from pale to rich pink. It is more tolerant of dryer soils than *A. perennis* but will still do best in moist soils that resemble those of its native wetland home.

Asclepias tuberosa, or butterfly milkweed, does best in upland locations but cannot tolerate wet feet. This species is most commonly sold in native and nonnative nurseries thanks to its tolerance of heat and drought, as well as its beaming orange blooms. It is well contained, growing only to around 2 feet tall, making it a perfect accent close to

Like many butterfly species, the male queen butterfly displays "hair penciling" behavior during courtship, in which he brushes his hair-pencils (brush-like organs that can be projected out from the body) against the antennae of the female, which secretes a pheromone that plays a significant role in successful seduction.

the home. Be careful not to confuse this native with the nonnative tropical milkweed, which typically presents with red and yellow blooms in addition to orange flowers.

While milkweed is best known for its use by monarchs, it is also the primary host plant of the queen (*Danaus gilippus*) and a secondary host plant by soldier (*Danaus eresimus*) butterflies.

Family: Asclepiadaceae (Milkweed Family)

Hardiness: Milkweeds are found throughout Florida, with distribution dependent on species; *A. tuberosa* and *A. perennis* grow in zones 8A–10B, with *A. incarnata* extending to 11

Lifespan: Short-lived perennials

Zone: Riparian–Upland, with *A. tuberosa* enjoying locations farther upland

Soil: Generally loam, sand, and clay, with variability being species dependent

Depth: *A. perennis* can tolerate up to 4 inches of water for extended periods of time, but, like *A. incarnata*, it requires bare soil to grow from seed; both do best in shoreline or seasonally inundated sites under 4 inches of water

Exposure: Species dependent

Growth Habit: Species dependent

Propagation: Seed

The dreamy flowers of *Conoclinium coelestinum* vary in color from lavender to a surreal blue.

MISTFLOWER
(Conoclinium coelestinum)

The heavenly mistflower is peculiarly named, and for good reason. When its blue, purple, or lavender flower clusters emerge, their protruding stamens of the same color give the plant a hazy appearance. Mistflower loves to spread throughout the garden, forming patches of almost triangular leaves. Throughout our state's moist meadows, riverine swamps, and swales, it grows wild, but it is also a fairly common wildflower for garden plantings. In addition to its lovely composition, the plant is attractive to many pollinating insects, especially skippers. In fact, almost all species in the sunflower family (Asteraceae) are important to pollinators and often are the mainstay of many nectar-drinking species, especially in dry, high disturbance, or areas of seasonal change.

This species is often not intentionally planted, as it enjoys disturbed sites and naturally recruits to community areas over time. However, many native nurseries enjoy cultivating it for its alluring aesthetic and pollinator benefits. Be cautious, though, when purchasing from box stores, as mistflower is often confused with *Praxelis clematidea*, a very similar-looking Category II invasive species.

Family: Asteraceae (Sunflower Family)
Hardiness: Zones 8A–11
Lifespan: Short-lived perennial
Zone: Riparian–Upland
Soil: Occasionally inundated to not extremely dry loam, muck, and sand

Depth: Shoreline, no permanent inundation, tolerates longer seasonal flooding
Exposure: Full sun to partial shade
Growth Habit: 1–2 feet tall
Propagation: Seed, division

Despite its name, the great purple hairstreak is actually quite blue.

With a high stress tolerance and abundant, cheery flower clusters, the narrowleaf yellowtop is one of the many native species that deserves more popularity in the home landscape.

NARROWLEAF YELLOWTOP

(Flaveria linearis)

The almost-always coastal, endemic narrowleaf yellowtop loves the wet prairies, pine rocklands, and disturbed habitats of our state. Outside of the sites it calls home, it is a hardy plant with a high stress tolerance, making it an ideal choice for more urban settings or locations where the soil has been disturbed. Luckily for onlookers, it blooms throughout fall and winter, dotting the landscape with relatively large groups of bright yellow flowers—hence the name. Depending on its maturity and site conditions, this golden inflorescence drowns out the narrow leaves and stems of the species, giving the appearance of floating, flaxen clouds.

Spreading wide and growing fairly low, it is an excellent option for the home landscape, where it will attract many species of butterfly and other pollinating insects. A somewhat surprising drought and salt tolerance lends this species to greater usage in coastal upland restoration, with Robinson Preserve in Manatee County being an excellent example of large-scale yellowtop restoration in mixed coastal community restoration projects.

Family: Asteraceae (Sunflower Family)
Hardiness: Zones 8A–11
Lifespan: Long-lived perennial
Zone: Riparian–Upland
Soil: Somewhat moist to dry sand, limerock

Depth: Shoreline, no permanent inundation, tolerates longer seasonal flooding, salt tolerant
Exposure: Full sun
Growth Habit: 2–3 feet tall and 3–4 feet wide
Propagation: Seed, division

As the name suggests, the eastern pygmy blue is the smallest butterfly on this side of the country.

Interestingly, an alternative common name for obedient flower is the very different eastern false dragonhead.

OBEDIENT FLOWER

(Physostegia purpurea)

Named for the manipulatable flower orientation on the stem, obedient flower is underutilized in the home landscape despite its charming blossoms and utility to pollinators. In the late spring through the beginning of fall, pink and purple tubular flowers with patterned throats emerge along erect racemes. This flush is attractive to long-tongued pollinators like hummingbirds and certain butterfly and bee species. While hummingbirds generally prefer the vibrantly red flowers of species like cardinal flower (*Lobelia cardinalis*), firebush (*Hamelia patens*), or coral honeysuckle (*Lonicera sempervirens*), they are exceptionally drawn to the obedient flower.

Throughout much of Florida, this wildflower is found in pinelands, rivers, and along marsh or swamp margins. Spreading via rhizome, the obedient flower will form relatively contained colonies—a plus for the hands-off gardener. Like many plants in the mint family, the obedient flower has square stems, an easy trick to aid identification in natural areas. Obedient flower is winter dormant in most areas, though it may persist in extreme South Florida.

Family: Lamiaceae (Mint Family)
Hardiness: Zones 8A–10B
Lifespan: Short-lived perennial
Zone: Riparian–Upland
Soil: Consistently wet to somewhat moist loam, muck, and sand

Depth: Shoreline, no permanent inundation, tolerates longer seasonal flooding
Exposure: Full sun to partial shade
Growth Habit: 3 feet tall
Propagation: Seed

Pickerelweed is perhaps the most notable of all the aquatic wildflowers.

PICKERELWEED
(Pontederia cordata)

Along with duck potato and knotted spikerush, pickerelweed makes up one-third of the pond shoreline trinity. From dense stands of bright green, lance-shaped leaves emerge stalks of tubular, vibrant purple flowers. While these stalks are individually short lived, the species grows in hearty patches, and during its blooming period (spring through fall) one will never be short an abundant crowd of blossoms. Rarely, the pickerelweed will also appear in pure white or light pink growth forms, which are occasionally commercially available.

Pickerelweed is a wildlife magnet! Ducks cannot get enough of it, making a treat of the entire plant. Its submerged portions are habitat for aquatic creatures and are known for providing excellent cover for sportfish like bass and, as the name suggests, pickerel. Pollinators of all sorts enjoy feasting on its consistently available nectar, and the seeds are snacked on by birds. Planting pickerelweed in a backyard pond or water garden is sure to make you a neighbor of all sorts of insects, birds, fish, and other creatures.

The whimsically named whirlabout is called such for the descending and ascending behavior of the adults, who take on a vortex-like, whirling flight pattern.

Beyond these qualities, the pickerelweed is also one of the best species to plant for shoreline stabilization and erosion control. Its intricate rhizomatic network holds sediment in place and provides a wave barrier. Even further, the plant filters and absorbs nutrients that, in the absence of these mechanisms, would potentially saturate the water body and lead to fish kills or algae blooms. The natural range in Florida includes the mainland but excludes the Keys. When planting, be sure to mix in other species as well to form a more mite- and disease-resistant colony.

Family: Pontederiaceae (Water Hyacinth Family)

Hardiness: Zones 8A–10B

Lifespan: Long-lived perennial

Zone: Littoral–Emergent

Soil: Aquatic to consistently wet muck and pond bottom

Depth: 3 feet is the maximum depth it can grow in long term, but it does best at 6 inches to 1 foot deep and can handle short dry periods

Exposure: Full sun to partial shade

Growth Habit: 3–4 feet tall

Propagation: Division

The saltmarsh mallow is a near twin of the swamp rosemallow.

SALTMARSH MALLOW

(Kosteletzkya virginica)

The saltmarsh mallow looks nearly identical to the swamp rosemallow, but take a closer look. *Kosteletzkya virginica* has a generally smaller flower and lacks the darker center that *Hibiscus grandiflorus* tends to have. In 1835, the *Kosteletzkya* genus was separated from *Hibiscus*, but it's clear to see why they were once one and the same. The flower of the saltmarsh mallow is made of very soft, pale to deep pink petals and bright yellow stamens. These flowers are edible raw or steeped into a tea and add a lovely botanical flair to a cake or salad. The leaves may also be eaten raw but are tough and therefore best when cooked down. Even the roots are edible—raw or cooked—and the water left over from boiling can be used as a substitute for egg whites. If that wasn't enough, its seed oil has been used as a biofuel, as well as for paints and varnishes, and is being considered for its use as an edible oil.

Beyond these practical uses, the plant is a wonderful addition to the aquatic butterfly garden, providing a reliable source of nectar from spring through fall. It is found growing wild in (you guessed it) saltmarshes, swamps, and coastal swales throughout most of the state.

Family: Malvaceae (Mallow Family)	**Depth:** Riparian
Hardiness: Zones 8A–10B	**Exposure:** Full sun to partial shade
Lifespan: Long-lived perennial	**Growth Habit:** 6 feet tall
Zone: Riparian–Upland	**Propagation:** Seed
Soil: Consistently wet to fairly moist muck and sand	

Even though it's not technically a pickleweed, the name would be accurate in reference to the briny taste of the succulent saltwort.

SALTWORT
(Batis maritima)

As its name suggests, saltwort can be eaten as a salty, salad herb and is relatively common in tropical and subtropical regions for that purpose. A sprawling groundcover with succulent foliage and tiny white flowers, it is a halophyte that wields its mat-forming powers for the good of coastline stabilization. It is also a common species in mangrove understories, and while not the showiest plant, it is helpful for filling in coastal areas. In these locations, it adds variety to what many feel are very simple landscapes.

It is the larval host for the southern white (*Ascia monuste*) and eastern pygmy blue (*Brephidium isophthalma*) butterflies. It is a close look-alike to the pickleweeds, or *Salicornia* genus, but is not related. This is an example of convergent evolution of species that enjoy salt flats, saltmarshes, and the margins of these coastal ecosystems.

The great southern white is the sole member of its genus *Ascia*.

| Family: Bataceae (Mustard Family) | Depth: Shoreline, no permanent inunda- |
| Hardiness: Zones 8B–11 | tion, tolerates daily saltwater flooding |

Family: Bataceae (Mustard Family)
Hardiness: Zones 8B–11
Lifespan: Long-lived perennial
Zone: Littoral–Riparian
Soil: Usually moist to occasionally inundated sand

Depth: Shoreline, no permanent inundation, tolerates daily saltwater flooding
Exposure: Full sun to partial shade
Growth Habit: 2–4 feet tall and many feet wide
Propagation: Cuttings

Although rather microscopic, the flowers of the sea purslane provide delicate sparks of animation amid its fleshy foliage.

SEA PURSLANE

(Sesuvium portulacastrum)

A vital dune stabilizer native to the entirety of Florida's coast, sea purslane is a visually delicate but ecologically vigorous perennial groundcover. Along beaches and marshes, it forms sprawling mats of fleshy, succulent leaves that are fully edible for humans. Like other edible coastal groundcovers, the sea purslane packs a salty punch, although this quality can be moderated with cooking. Since it is native to tropical and temperate coastlines worldwide, it is widely grown and consumed, although wild collection is discouraged to protect dune ecosystems. Mote Aquaculture Research Park has been experimenting with growing sea purslane and other brackish and salt-tolerant plants in mixed aquaponics systems to test their potential value for filtering water in fishery systems. The crop is high in antioxidants, vitamins, and beta-carotene.

Blooming throughout the year, sea purslane puts forth tiny, brightly pink, star-shaped flowers that are a known nectaring source for the rare Miami blue (*Clyclargus thomasi*). These blossoms open and close in a single day. It is an ideal groundcover for the sunny coast or simply to expand your edible and medicinal plant collection.

Family: Aizoaceae (Fig-Marigold Family)
Hardiness: Zones 8A–11
Lifespan: Long-lived perennial
Zone: Riparian–Upland
Soil: Aquatic to not extremely dry sand

Depth: Shoreline to several inches, doing best with tidal fluctuations
Exposure: Full sun
Growth Habit: 6 inches tall
Propagation: Division

No need to fear! The small-spike false nettle is soft to the touch and doesn't possess any stinging abilities.

SMALL-SPIKE FALSE NETTLE
(Boehmeria cylindrica)

While it looks like the infamous stinging nettle (*Urtica diocia*), no need to fear—the false nettle (also referred to as bog hemp) lacks the perilous hairs that give stinging nettle its name. Native to all counties in the state and found in habitats like cypress swamps, marshes, wet forests, and ditches, *Boehmeria cylindrica* is a great choice for restoration in and around ponds, lakes, and bioswales. Its flowers grow from the axils of its upper leaves in a cylindrical shape and are greenish white in color. While these blossoms do not attract many insects, the plant is indeed a larval host for eastern comma (*Polygonia comma*), question mark (*Polygonia interrogationis*), and red admiral (*Vanessa atalanta*) butterflies, who feed on its ovate, classically green foliage.

While this species is a rather unlikely find in nurseries, don't let that *bog* you down. Bog hemp often naturally recruits into wetland gardens. Seed is an excellent way to help start this species in your landscape.

Despite the name, bog hemp is not a true cannabis of the Cannabaceae family, but rather a nettle of the Urticaceae family.

Family: Urticaceae (Nettle Family)
Hardiness: Zones 8B–11
Lifespan: Short-lived perennial
Zone: Riparian
Soil: Aquatic to occasionally inundated clay, loam, and muck

Depth: Shoreline, no permanent inundation, tolerates longer seasonal flooding
Exposure: Partial shade to full shade
Growth Habit: 2–4 feet tall and 1–3 feet wide
Propagation: Seed, root cuttings

Red admirals are highly territorial, with males constantly patrolling their area and chasing off intruders. Females mate only with territory-holding males.

The sunny blooms of the southeastern sneezeweed add a jubilant spirit to the garden.

SNEEZEWEEDS
(Helenium spp.)

There are seven species of the deceivingly named sneezeweed native to Florida: *H. bifidum*, *H. amarum*, *H. flexuosum*, *H. pinnatifidum*, *H. autumnale*, *H. brevifolium*, and *H. vernale*. Of these, common sneezeweed (*H. autumnale*), southeastern sneezeweed (*H. pinnatifidum*), and bitter sneezeweed (*H. amarum*) are the most commonly found (either in pots or in seed packets) at native nurseries.

Common sneezeweed is found growing wild in all sorts of wetland ecosystems, from floodplain forests to bogs. A single individual can produce a hundred blooms, growing anywhere from 2 to 4 feet tall. Like other sneezeweeds and similar flowering perennials, this species can be cut back to form a bushier specimen or allowed to grow freely, winding upward on taller stalks. The southeastern sneezeweed grows slightly smaller, reaching heights of only around 3 feet. Like the common sneezeweed, southeastern sneezeweed loves moist to wet soils and full sun conditions. Bitter sneezeweed is significantly less common commercially, but a great, drought-tolerant choice for upland locations. Its foliage is fine and branching, creating a unique and lovely display when allowed to reseed in open locations.

The genus *Helenium*, most sources state, refers to Helen of Troy, the Greek god Zeus's daughter and legendary cause of the Trojan War. Myth says these flowers rose from the ground where her tears fell. Since this plant was only found in Europe as recently as 1729, it's safe to say this story takes some botanical liberties.

Family: Asteraceae (Sunflower Family)
Hardiness: 8A–10B
Lifespan: Long-lived perennial
Zone: Upland
Soil: Typically moist clay, loam, and sand

Depth: Shoreline, no permanent inundation, tolerates short periods of flooding
Exposure: Full sun
Growth Habit: 2–4 feet depending on species
Propagation: Seed

The string lily may not technically be a lily, but it's certainly as delightful as one.

STRING LILY

(Crinum americanum)

Native to wet prairies, riverine swamps, and similar aquatic habitats throughout most of Florida, the string lily is a jewel of the waterside. Evergreen and erect, the plant (which is not actually a member of the lily family) reveals its large, white flowers at almost any time of the year, although typically in the spring for northern counties and later in the year farther south. These flowers are aromatic and attractive to many pollinators, particularly sphinx moths. Its bulbs, though poisonous to humans and other mammals, are a favored snack of the lubber grasshopper.

Oddly, while string lilies (also known as crinum lilies) can live for extended periods in several inches of water, tolerate the flow of rivers, grow in floating mats, and even withstand estuary influences, their maximum depth is variable. When water is above where their leaves split from their stalk for long periods, especially in environments with high water flow, they'll often break off at their surface roots and "go with the flow." When water levels lower, or they're pushed onto more favorable muck, they will reroot. Size, sun exposure, the presence of other lilies nearby, depth, flow, and seasonal changes all affect this behavior. The authors would like to note that this behavior is most common in rivers with new transplants, smaller specimens, or on eroding pond banks.

For best results in the aquatic garden, start shallower than you might think, as they can tolerate moist upland conditions and will let their rhizomes grow farther into the depths as they'd like. Despite being rhizomatous, they are poor at controlling erosion—choose a grass, tree, or shrub species instead.

Family: Amaryllidaceae (Amaryllis Family)
Hardiness: Zones 8A–11
Lifespan: Long-lived perennial
Zone: Riparian–Upland
Soil: Aquatic to somewhat moist dry loam, muck, and sand

Depth: Shoreline with permanent inundation, tolerates longer seasonal flooding, tolerates tidal flooding; unique floating and root-breaking behavior in some circumstances
Exposure: Full sun to full shade
Growth Habit: 1–2 feet tall
Propagation: Seed, bulb division

The colony-forming habit of swamp sunflower is much valued by those who appreciate its jovial blooms.

SWAMP SUNFLOWER
(Helianthus angustifolius)

One of the most common sunflowers native to Florida, the swamp sunflower forms dense colonies in our wetland and moist meadow ecosystems. When autumn hits, it puts on a spectacular display of star-shaped golden blooms. In locations where the plant has grown wild over time, these flushes are breathtaking. Due to this colony-forming habit, gardeners may want to avoid planting in locations where the swamp sunflower may outcompete other wildflowers or otherwise extend beyond its desired range. As it is taller than most other wildflowers, it is useful in road ditches to add vibrancy to open space or behind other flower displays to create a lively backdrop.

The species is primarily pollinated by bees, although it is visited by plenty of other pollinating insects. Birds and small mammals enjoy the seeds, and creatures of all sorts take cover in its meadowed stands. Like many sunflowers, swamp sunflowers are able to sequester some pollutants and heavy metals at a higher rate than most plants. This characteristic makes them especially useful in superfund sites, parking lot stormwater systems, and areas where car runoff in particular can be an issue.

Family: Asteraceae (Sunflower Family)
Hardiness: Zones 8A–10A
Lifespan: Short-lived perennial
Zone: Riparian
Soil: Consistently wet to fairly moist loam and sand

Depth: Shoreline, no permanent inundation, tolerates longer seasonal flooding
Exposure: Full sun
Growth Habit: 5–8 feet tall
Propagation: Seed, division

The thistle may have an ornery body, but its head is quite the opposite—gentle and darling.

THISTLES

(Cirsium spp.)

The seven native thistles of Florida are known far more for their prickly defenses than their benefits to pollinators. For those more familiar, the thistle rightfully claims a place in the garden—even if precautions might be necessary to keep its formidable defenses far from straying hands and feet.

Ironically, the horrid thistle (a name that certainly doesn't do the plant any favors in the eyes of the public) is the most popularly cultivated of our native *Cirsium* species. This wildflower winters as a basal rosette in our state's coastal plains, saltmarsh edges, pinewoods, and disturbed areas, erecting an impressively spiked stalk from which blossoms a lovely purple flower. This bloom, almost surprisingly delicate in contrast with the threatening silhouette of the rest of the plant, can emerge in shades from lavender to vibrant violet. After flowering, the head of purple is replaced by cotton-like seed heads that ensure the thistle's persisting presence in the landscape. These seeds contain oil that makes them a highly valued food source for birds. The thistle is also the host plant for the painted lady (*Vanessa cardui*) and little metalmark (*Calephelis virginiensis*) butterflies and is otherwise an immensely favored nectar source for a wide assortment of others, especially swallowtails.

Not every home landscape will be equipped to host such a host plant, but those who are willing to take up the torch will be rewarded with plenty of pollinator activity. The plant is perfect for locations the gardener seeks to truly rewild, letting their native garden naturalize and develop freely. The thistle's defenses can also be used strategically in the garden, as planting the species as a barrier to surround more precious species will prevent herbivory and similar disturbances. While the horrid thistle is the most likely to be found potted in your local native nursery, more moisture-loving species like the swamp thistle (*Cirsium muticum*) are available online from trusted seed distributors.

Family: Asteraceae (Sunflower Family)
Hardiness: 8A–11
Lifespan: Biennial
Zone: Upland
Soil: Somewhat moist to extremely dry sand for *C. horridulum*

Depth: No permanent inundation, tolerates shorter seasonal flooding, not salt tolerant
Exposure: Full sun to partial shade
Growth Habit: 1–4 feet depending on species
Propagation: Seed

The tickseed is featured on the State Wildflower license tag, the only consistent source of funding for Florida's native plants and wildflowers. The tag has raised more than $4.2 million since its conception, supporting everything from school wildflower education gardens to plantings of wildflower "pollinator pathways" along roadsides.

TICKSEEDS

(Coreopsis spp.)

Florida doesn't have just one state wildflower—it has twelve! The twelve native species from the genus *Coreopsis*, or tickseeds, are the sunny, blossoming representatives of our state. These flowers come in varying colors, growth habits, and environmental preferences, although most resemble daisies with a ring of yellow petals surrounding a yellow or brown center. Many of the twelve species are found almost exclusively in northern counties, although the most common, Leavenworth's tickseed (*Coreopsis leavenworthii*), is found throughout the entirety of the state. This species is native to the mesic pine flatwoods but is more often found dotting roadsides and highway medians—and, of course, in butterfly gardens! Tickseeds attract many pollinators to their vibrant flushes, with colors ranging from a sublime mauve to an ombre crimson. Most tickseeds are annuals or short-lived perennials that vigorously reseed—a welcome trait in landscapes that are suitable for bountiful settlements of the cheery native.

Tickseeds vary in growth habit, with some being significantly bigger than others. *C. floridana* is standout in this manner, with a similar size to *C. leavenworthii* but with much larger flowers. *C. floridana* is endemic to the state and occupies similar areas to *C. leavenworthii*. Both species tolerate longer periods of flooding but need open, dry ground for germination to occur.

Family: Asteraceae (Sunflower Family)
Hardiness: 8A–11
Lifespan: Annual, but year-round growth
Zone: Riparian–Upland
Soil: Species dependent
Depth: *C. leavenworthii* and *C. floridana* both tolerate longer seasonal flooding and can grow long term in water or floating weed mats but need dry, bare ground to seed properly
Exposure: Full sun
Growth Habit: 1–4 feet depending on species
Propagation: Seed

Water hyssop may typically grow only up to 1–2 inches, but it sprawls several feet, bringing erosion control and pollinator benefits wherever it goes.

WATER HYSSOP

(Bacopa monnieri)

A nonaromatic, creeping herb, water hyssop is a fantastic groundcover that attracts small and low-lying pollinators with its subtle whitish lavender flowers that adorn trailing stems. It finds its home along lake and pond edges, swamps, marshes, and similar habitats where it helps stabilize shorelines and prevent erosion with its often dense network of growth. It is the larval host for the white peacock butterfly (*Anartia jatrophae*). *Bacopa monnieri* is a dominant and widespread groundcover found in numerous wetland, aquatic, or perpetually moist ecosystems.

Water hyssop (also known as brahmi or herb-of-grace) is used in Ayurvedic traditional medicine in the belief that it improves memory and other cognitive functions. Recent studies have affirmed these uses and more, pointing to the compounds within as having unique pharmacological potential.

Lemon bacopa (*Bacopa caroliniana*) is a relative that boasts blue flowers and is shorter lived and less tolerant of drier conditions. *B. caroliniana* is less widespread, with the authors seeing it most often in sandy bogs or sparsely grown spikerush wetlands in central Florida.

Family: Plantaginaceae (Plantain Family)

Hardiness: Zones 8B–10B

Lifespan: Long-lived perennial

Zone: Littoral–Riparian

Soil: Aquatic to usually moist muck or sand

Depth: As long as light can still penetrate, *Bacopa* species can grow up to several feet, but they generally prefer shoreline areas that are seasonally exposed where they can behave like a traditional groundcover

Exposure: Full sun to partial shade

Growth Habit: 6 inches tall and several feet wide

Propagation: Seed, cuttings, and division, with the latter being the best method

While the males and females don't differ in appearance, white peacock butterflies express seasonal polyphenism, in which they are smaller and darker during the wet season of summer and paler and larger in the dry winter.

While it may not possess the popular blue appearance of the nonnative plumbago, the wild plumbago is definitely worthy of its own praise.

WILD PLUMBAGO
(Plumbago zeylanica)

The nonnative blue plumbago gets most of the attention, but our homegrown wild plumbago is just as practically valuable to the home landscape, sprinkling the understory with phlox-like, aromatic white flowers. These blooms extend from the draping, rounded form of the plumbago, which retains its evergreen nature in counties without hard freezes.

Wild plumbago serves as the larval host plant for the dainty cassius blue butterfly (*Leptotes cassius*) and tolerates shade like a champ. It can grow and sprawl rather quickly but will handle pruning well. This species grows wild in our state's coastal hammocks. Wild plumbago is an excellent understory groundcover, doing well in back areas to fill space beneath trees, with its flowers coming forward toward more sunlit areas.

Family: Plumbaginaceae (Leadwort or Plumbago Family)
Hardiness: Zones 9A–11
Lifespan: Long-lived perennial
Zone: Upland

Soil: Occasionally inundated to very dry lime-rock and sand
Depth: Shoreline only
Exposure: Full sun to full shade
Growth Habit: 1 foot tall and 3–4 feet wide
Propagation: Seed

The Simpson's zephyr lily may not look like much when not in bloom, but the plant provides a soft allure whenever it graces us with its presence.

ZEPHYR LILIES
(Zephyranthes spp.)

The Simpson's zephyr lily, *Zephyranthes simpsonii*, is unassuming when not in bloom. Its foliage is fine, slender, and muted green in color, blending seamlessly into natural landscapes. In the springtime, humble stalks rise to present the softly white, ephemeral blooms. These flushes will often push forth after a rainfall event, giving the plant its other, less-used designation of simply "rainlily."

Northern residents should consider *Zephyranthes atamasca* its equivalent. Zephyr lilies are relatively rare and usually found in open, wet grasslands, dome swamps, and moist areas without too much shade. Fire promotes the growth of these species, as well as mowing or heavy grazing, making them a feature of many cow pastures and roadsides. While they are fairly uncommon in nurseries due to their small foliage and ephemeral blooms, they are in cultivation in choice locations. Experienced gardeners tend to use these as specimen plantings in front areas of beds, as they are low to the ground even in bloom.

Family: Amaryllidaceae (Amaryllis Family)
Hardiness: Zones 9A–10B
Lifespan: Long-lived perennial
Zone: Upland
Soil: Somewhat moist to dry sand

Depth: No standing water—only moist, well-draining locations
Exposure: Full sun to partial shade
Growth Habit: 1 foot tall
Propagation: Seed, bulb division

The cassius blue cannot survive even the winters of northern Florida but thrives on the warm coasts of the peninsula.

While grasses are not known for their aesthetic beauty, the textured poise of the bushy blue-stem should make it a valued adornment to the landscape.

4

GRASSES

BUSHY BLUESTEM

(Andropogon glomeratus)

In any landscape, it's often the lush wildflowers that receive the most accolades. We associate these flowers with beauty—their subtleties and intricacy fascinate our eyes, minds, and hearts as we gaze on them. Perhaps, then, we should take a moment to extend this view beyond where it typically focuses: into the world of grasses and rushes. In this world, natives like the bushy bluestem flower in a different but equally enchanting way.

Native to flatwoods and marshes, the bushy bluestem forms stands of handsome, erect stalks of silvery green. As these protrusions age into autumn, they turn a curious copper tone that stands out against a more verdant backdrop. Who said Florida doesn't have a fall?

A male Georgia satyr, usually on patrol for mates, finds success with the reclusive female.

The flowers of the bushy bluestem are its most recognizable feature. These dense tufts emerge like cotton candy, visually softening the habitats it calls home. The genus name *Andropogon* refers to the Greek words *aner* and *andros* (meaning "man") and *pogon* (meaning "beard"). Gazing on grasses of the genus, it is easy to see why. Its seeds are eaten by birds and small mammals and are important for songbirds in the winter. It is occasionally browsed by deer and provides nesting material for birds.

Bushy bluestem has several subspecies with varying affinities for wetland conditions. *Andropogon glomeratus* var. *glaucopsis* is sometimes considered a separate species and is distinct for its bluish purple coloration. *Andropogon glomeratus* var. *pumilus* is the most widely distributed throughout Florida and likely the most common. It is the larval host plant for the following butterflies, many of which are grass generalists: Delaware skipper (*Anatrytone logan*), neamathla skipper (*Nastra neamathla*), swarthy skipper (*Nastra lherminier*), and twin-spot skipper (*Oligoria maculata*).

Family: Poaceae (Grass Family)
Hardiness: Zone 8A–11
Lifespan: Long-lived perennial
Zone: Riparian–Upland
Soil: Consistently wet to not extremely dry loam, muck, or sand

Depth: Shoreline only but tolerates seasonally flooding less than 6 inches
Exposure: Full sun
Growth Habit: 3–5 feet tall and 1–3 feet wide
Propagation: Seed, division, sprigging

Elliott's lovegrass stands out in the verdant landscape with some-what silvery, even pinkish leaves.

ELLIOTT'S LOVEGRASS
(Eragrostis elliottii)

With a name as sweet as lovegrass, it is no wonder that this native is such a delight. This clump-forming species grows throughout Florida in flatwoods, prairies, sandhills, and disturbed sites. Wherever it may be, it adds welcome diversity in visual texture to the landscape with its tiny blooms, which flower in such abundance that the entire plant takes on an undulating, foggy beige appearance. These tend to appear in autumn but, depending on site conditions, may blossom throughout the year.

Its prolific seeds provide a reliable food source for all sorts of creatures, from invertebrates to birds, who use the hazy grass for cover. Elliott's lovegrass is the larval host for the zabulon skipper butterfly (*Poanes zabulon*). Once established, it is fairly drought tolerant and can survive occasional inundation of fresh *and* brackish water. As a cherry on top, the grass also lends a hand in moderating erosion.

For a showier relative in drier, less salty areas, try using the purple lovegrass (*Eragrostis spectabilis*), which looks very similar but has a slightly more purple coloration.

Family: Poaceae (Grass Family)
Hardiness: Zone 8A–10B
Lifespan: Short-lived perennial
Zone: Riparian–Upland
Soil: Occasionally inundated to not extremely dry loam and sand

Depth: Shoreline only but can tolerate light seasonal flooding
Exposure: Full sun
Growth Habit: 1–3 feet tall and 1–2 feet wide
Propagation: Seed, division

The zabulon skipper used to be called the southern golden skipper. Unlike many skippers, it has a range of patterning on the wings and exhibits sexual dimorphism, making identification difficult.

The flowers of the fakahatchee truly rival that of many wild-flowers.

FAKAHATCHEE GRASS
(Tripsacum dactyloides)

There's nothing like a stand of fakahatchee grass. Evergreen and clump forming, this native (which is also commonly referred to as eastern gamagrass) is commonly planted in a continuous line to delineate and refine walkways and other structures. Its "model" growth form and hearty look make it a great go-to in the home landscape, and its ecological value only adds to its allure. It's hard, yellow seeds are consumed by deer and birds, and its rhizomatous roots are a line of defense against erosion.

Flatwoods, wet bogs and hammocks, and river banks are just a few of the numerous ecosystem types where fakahatchee grass can be found growing wild. Unlike many of its grass family relatives, its flowers are unusually flamboyant, draping in rows along extended, nodding spikes that rise above the leaves. These warm-toned flowers are often bold in color, ranging from pink to rust. Gardeners will be happy to know the grass is moderately drought tolerant and highly tolerant of flooding. It can be cut back or pruned to one's liking but ultimately requires little to no maintenance. A note of caution: The leaves have very small, sharp teeth along their margins, which can cause an unexpected cut to those handling the plant without gloves.

A dwarf variety, referred to as *Tripsacum floridanum*, grows only 2–3 feet tall and wide. Otherwise, the species are nearly identical and easily confused. Both are host plants for the byssus skipper butterfly (*Problema byssus*). Only the full-sized fakahatchee has been shown to serve as a host for the three-spotted skipper (*Cymaenes tripunctus*) and clouded skipper (*Lerema accius*). However, like many grasses, these species are likely hosts for many generalist skipper butterflies.

Family: Poaceae (Grass Family)
Hardiness: Zone 8A–10B
Lifespan: Long-lived perennial
Zone: Riparian–Upland
Soil: Occasionally inundated to not extremely dry loam and sand

Depth: Shoreline with seasonal freshwater inundation
Exposure: Full sun to partial shade
Growth Habit: 4–6 feet tall and 2–4 feet wide
Propagation: Seed

The three-spotted skipper is named for its three translucent spots located near the tip of its forewing. These are generally quite difficult to see, leading one to wonder whether another common name might more efficiently suffice.

The dense nature of maidencane is clearly seen here.

MAIDENCANE
(Eragrostis elliottii)

Maidencane is assertive—so much so, in fact, that it has a tendency to form an eco-system type of its own: the "maidencane marsh," a rooted or floating monotypic stand thriving off the plant's exceptionally thick network of rhizomes. Common in all sorts of freshwater wetlands, this keystone species is a plant highly utilized in restoration. In locations where its vigorous spread is not welcome, it can be considered a weed, but the maidencane is simply doing what it was born to do: establish thick and quick, pro-viding cover and food for aquatic organisms, small mammals, birds, deer, the American alligator, and even the Florida panther.

In addition to serving nearly every creature that roams the marshlands, the maid-encane is the larval host for the clouded skipper (*Lerema accius*) and Delaware skipper (*Anatrytone logan*) butterflies. Due to its aggressive growth habit, the maidencane isn't recommended for home landscapes where it would need to be contained.

Family: Poaceae (Grass Family)
Hardiness: Zone 8A–10B
Lifespan: Long-lived perennial
Zone: Littoral–Riparian
Soil: Consistently wet to somewhat moist muck, sand, and pond bottom

Depth: Maidencane tolerates permanent inundation of several feet if connected to shallower areas
Exposure: Full sun to partial shade
Growth Habit: 3–4 feet tall
Propagation: Seed, division

Switchgrass was the king of the prairie prior to European colonization.

SWITCHGRASS
(Panicum virgatum)

Before the European colonizers began their trek across the continent, converting prairies to monocrops, much of our country was home to spacious, rolling habitats dominated by native grasses. Switchgrass was one of those that dominated and is still valued today for the same resourceful and rugged qualities that made it a prime feature of the presettled landscape. Native to the flatwoods, marshes, riverine forests, and disturbed sites of our state, the clump-forming switchgrass displays light green foliage that yellows in autumn. Its seeds are eaten by birds, who also use it for cover and nesting material.

Throughout the country, switchgrass is used as a groundcover for soil conservation, as biofuel, for grazing and foraging, to control erosion, and even as a substitute for wheat straw for livestock bedding and mushroom-growing substrate. It is also the larval host for the tawny-edged skipper (*Polites themistocles*), Delaware skipper (*Anatrytone logan*), and dotted skipper (*Hesperia attalus*) butterflies.

Family: Poaceae (Grass Family)
Hardiness: Zone 8A–11
Lifespan: Long-lived perennial
Zone: Riparian–Upland
Soil: Occasionally inundated to somewhat moist loam and sand

Depth: Shoreline only but can tolerate brief seasonal flooding
Exposure: Full sun to partial shade
Growth Habit: 3–4 feet tall and 2–3 feet wide
Propagation: Seed, division

Clouded skippers are generalists that will readily consume nearly any broad-bladed grass—even lemongrass, an intensely flavored herb filled with all sorts of compounds that are typically distasteful to insects.

In the first half of the twentieth century, the tawny-edged skipper was regarded as one of the most common and widely found skippers on this side of the country, but due to destruction and fragmentation of meadow habitat, it is significantly less common, though not threatened.

Coming across the American wisteria in bloom for the first time is not an event a naturalist forgets.

5

VINES

AMERICAN WISTERIA
(Wisteria frutescens)

Wisteria is a name many won't forget once they see the plant it references. While the invasive *Wisteria* species native to Asia are the most common throughout the United States, we have a native with flowers just as worthy of popularity: the American wisteria! In form, the large, woody vine is similar to the prized but nonnative bougainvillea (*Bougainvillea* spp.) but features humongous, draping clusters of large, purple flowers. The tone of these flowers varies from lavender to deep amethyst, and white morphs exist and are found occasionally in cultivation.

American wisteria grows only as far south as zone 9B and is mostly found in the Panhandle, taking up residence in floodplains and uplands thickets. The species is a host plant for the silver-spotted skipper (*Epargyreus clarus*) and long-tailed skipper (*Urbanus proteus*). A plethora of pollinators enjoy its flowers, and the foliage is often eaten by deer and other browsers. It is slower growing, reducing maintenance costs in the long term, but do not expect smaller plants or seedlings to reach full size overnight. It is gorgeous on a trellis, as a shade canopy, or to add color to the base of trees and buildings in more open areas where climbing can be utilized.

The authors have documented the species in the wild alongside Florida anise (*Illicium floridanum*), buckwheat tree (*Cliftonia monophylla*), southern magnolia (*Magnolia grandiflora*), red maple (*Acer rubrum*), and American sweetgum (*Liquidambar styraciflua*), all of which make a great pairing in the home landscape.

Family: Fabaceae (Legume Family)
Hardiness: Zone 8A–9B
Lifespan: Long-lived perennial
Zone: Upland
Soil: Loam, sand

Depth: Shoreline only but tolerates seasonal flooding
Exposure: Full sun to partial shade
Growth Habit: 30 feet tall and 4–8 feet wide, woody base
Propagation: Seed, cuttings

The vibrant tubular flowers of the coral honeysuckle are an iconic part of the tropical landscape.

CORAL HONEYSUCKLE

(Lonicera sempervirens)

Of the most notable flowering plants native to Florida, coral honeysuckle makes a name for itself with its iconic orangey pink to red tubular flowers. The vine is commonly cultivated by pollinator enthusiasts who cherish witnessing the long-tongued butterflies and hummingbirds that flock to its blooms. Luckily for the gardener as well, this plant will swallow any structure on which it is placed if conditions are right. Fences on which it is grown readily transform into vibrant, tropical hedges. In most counties, the vine is evergreen, but it will become deciduous farther north. Farther south, it will bloom reliably year-round and is considered a critical part of ruby-throated hummingbird (*Archilochus colubris*) migration into Florida, alongside other red tubular flowers such as firebush (*Hamelia patens*), coral bean (*Erythrina herbacea*), and crossvine (*Bignonia capreolata*).

Coral honeysuckle is also the larval host of the spring azure (*Celastrina ladon*) and snowberry clearwing (*Hemaris diffinis*) butterflies. Although this plant can be found in big box stores, it is recommended to restrict purchasing to native nurseries, as many of these larger stores use pesticides that kill caterpillars.

Some vines make a name for themselves by being destructive to buildings or residential landscapes due to sticking feet or powerful, crushing branches and roots. Coral honeysuckle is not one of these, and while this quality means your paint is safe, it will have trouble climbing slicker structures, like windows.

Family: Caprifoliaceae (Honeysuckle Family)

Hardiness: Zone 8A–10B

Lifespan: Long-lived perennial

Zone: Upland

Soil: Occasionally inundated to very dry loam, muck, or sand

Depth: Shoreline, no permanent inundation but can tolerate very mild seasonal flooding

Exposure: Full sun to full shade

Growth Habit: 10–15 feet on average

Propagation: Seeds, cuttings

The unparalleled flower of the maypop passionvine is one that will never be forgotten once spotted.

MAYPOP PASSIONVINE

(Passiflora incarnata)

It is hard to gaze on the flower of the maypop passionvine and not get lost in its intricate, cosmic bloom. Its ten petals, purple-crowned corona, and pink filaments inspired Christian missionaries to declare it the "passionflower," associating each of its conspicuous features symbolically with elements of the storied death of Jesus Christ. It is easy to see how the passionflower could be so moving. Not only does it stun with such an incomparable bloom, but its edible fruits are also a delicacy. The passionfruit's tough rind encases a membranous, seed-filled sac of sweet and tangy pulp that is used across the world in desserts from pavlova to cheesecake.

Julia heliconian butterflies have been observed to agitate the eyes of turtles in order to drink their tears.

The perennial vine is known among gardeners for its fast-growing, aggressive establishment. Wherever you place maypop, be aware that it will extend its reach as far as possible, especially when receiving full sun and summer rains. It is drought tolerant and moderately salt tolerant and will die back in winter. While the average size is only around 20 feet, it can grow larger under certain circumstances. Maypops can easily reach the top of oak canopies and drape down when given the option. To protect young plants or to ensure even coverage when on a trellis, we recommend mixing it with other vines.

The maypop passionvine is a host plant for the gulf fritillary (*Agraulis vanillae*), variegated fritillary (*Papilio glaucus*), zebra longwing (*Heliconius charithonia*), crimson-patched longwing (*Heliconius erato*), Julia heliconian (*Dryus iulia*), banded hairstreak (*Satyrium calanus*), and red-banded hairstreak (*Calycopis cecrops*) butterflies. Interestingly, sun exposure seems to alter the butterfly species attracted to the plant. Full sun tends to attract more gulf fritillaries, while shadier locations are more frequented by zebra longwings.

Family: Passifloraceae (Passionflower Family)
Hardiness: Zone 8A–10B
Lifespan: Short-lived perennial
Zone: Upland
Soil: Somewhat moist to extremely dry clay and sand

Depth: Shoreline only but can adapt to seasonal shallow flooding of several inches
Exposure: Full sun to partial shade
Growth Habit: 25 feet or more
Propagation: Seed, cuttings

The wooly pipevine is a handsome feature of the landscape, even if a little strange.

PIPEVINES

(Aristolochia spp.)

Pipevines are a staple in the world of butterfly gardening, with three species found in Florida. Pipevines are defined by their handsomely curving flowers resembling smoking pipes, usually presenting with deep burgundy and purple highlights with heavy venation. They serve as host plants for the beautiful pipevine swallowtail (*Battus philenor*) and the endangered polydamas swallowtail (*Battus polydamas*).

The wooly dutchman's pipevine (*Aristolochia tomentosa*) is the most water tolerant and second most common of the three native pipevines. Its drooping flowers have yellow and purple entrances, and its thick, heart-shaped leaves make a wonderful backdrop on trellises or mixed with other vines in the home landscape. The dutch-man's pipevine (*A. macrophylla*) is the most recognizable of the three species and is slightly larger than the wooly pipevine, in both size and flowering. It is widespread in states farther north but fairly uncommon in Florida, limited to only the westernmost northern counties. Virginia snakeroot (*A. serpentaria*) is the odd one out of the three, reaching only 3 feet long at maximum, but it is a fun groundcover curiosity if locally available. All pipevines native to the United States are threatened to some degree in at least one state. For this reason, we recommend purchasing only from reputable native plant growers.

Several nonnative pipevines, such as the commonly planted calico vine (*A. littora-lis*), are considered invasive, and they also actively displace native species and show evidence of being toxic to native pipevine swallowtails, reducing their numbers over time.

Family: Aristolochiaceae (Birthwort Family)
Hardiness: Zone 8A–8B
Lifespan: Long-lived perennial
Zone: Upland
Soil: Occasionally inundated to somewhat moist loam or sand

Depth: Shoreline, no permanent inundation; wooly pipevine tolerates light seasonal flooding
Exposure: Full sun to partial shade for the wooly pipevine
Growth Habit: 20 feet or more for the wooly pipevine
Propagation: Seed

The swamp leather flower, while not commonly planted in the home landscape, is ideal for somewhat awkward areas such as stairs and low walls.

SWAMP LEATHER FLOWER

(Clematis crispa)

Swamp leather flower takes a vining form among the wet woods and marshes of Florida. During its long blooming period from spring through the fall, it displays its pinkish purple, bell-shaped flowers that bow softly along the stem. These flowers attract a range of pollinators, including hummingbirds! The seeds also provide food for birds and other wildlife.

While this vine is suitable for a trellis and similar vertical structures that will encourage its climb, its habit and fairytale flower structure make it a particularly apt candidate for growing along formations like stairs and low walls that lack support. It is usually not a dominant vine and is best used as an accent species. All *Clematis* species in Florida are host plants for the mournful thyris (*Thyris sepulchralis*), a moth that serves as a great example of a colorful, day moth that thrives in tropical areas around the world. Netleaf clematis (*Clematis reticulata*) is a very drought-tolerant relative of the swamp leather flower, and virgin's bower (*Clematis virginiana*) falls moderately between the two species in terms of moisture tolerance.

Family: Rannunculaceae (Buttercup Family)
Hardiness: Zone 8A–9B
Lifespan: Long-lived perennial
Zone: Riparian–Upland
Soil: Occasionally inundated to not extremely dry loam, muck, or sand

Depth: Shoreline only, but often hangs over shaded rivers and wetlands
Exposure: Partial shade
Growth Habit: 4–10 feet
Propagation: Cuttings

GLOSSARY

algae: A highly diverse group of photosynthetic, eukaryotic organisms that are most often aquatic. Single-celled organisms like *Karenia brevis* (commonly referred to as "red tide") are algae, as are many of the species we call seaweeds. In the field of aquatic management, we tend to be referring to two main visually identifiable forms: filamentous mats that are thick and hairlike, and planktonic, which is microscopic and turns the water a green, red, or brown hue.

allelopathy: A biological phenomenon in which a plant releases chemicals that impact the life cycle of other plants. For example, the fallen leaves of the hackberry (*Celtis occidentalis*) release chemicals called phenolic phytotoxins that inhibit the germination and development of seedlings, especially grasses.

annual: A plant that completes its life cycle within one year, from germination to seed.

anther: The part of a stamen that contains the pollen.

basal: Pertaining to the base of a plant or its leaves, often referring to growth at the base.

biennial: A plant that takes two years to complete its life cycle.

bog: A type of ecosystem that accumulates peat (a soil-like accumulation of partially decayed vegetation) and is characterized by acidic waters and low nutrient levels.

catkin: A cylindrical flower cluster that is often pendulous.

clonal: Clonal refers to the asexual reproduction of plants that are genetically identical to their parent, from which they've emerged vegetatively. Many of the most invasive species worldwide are clonal. Clonality is a complex ecological phenomenon whose impacts on ecosystem health and function are contextual and nuanced.

coastal plain: A type of ecosystem characterized by its flat, low-lying nature and being adjacent to the ocean.

coevolution: The reciprocal influence of species on each other's evolution.

convergent evolution: The process whereby organisms independently evolve similar traits, typically as a result of adapting to similar environments.

corm: A fleshy underground plant stem.

cultivar: A variety of plant that has been produced in cultivation by selective breeding. The horticultural practice of such selection is fundamental to the commercial plant industry resulting in everything from heat-tolerant varieties of an otherwise heat-sensitive plant to multiple color options of the same species.

cuticle: A hydrophobic, protective, extracellular layer covering the surface of a plant.

cycad: An ancient group of dioecious, cone-bearing plants that used to dominate the planet. The coontie (*Zamia integrifolia*) is the only cycad native to Florida.

deciduous: A term used to describe plants that lose their leaves seasonally.

dioecious: Having male and female reproductive organs on separate plants.

ecotype: A distinct form of a plant species that has adapted to specific environmental conditions.

emergent: Referring to plants that grow in water but whose leaves and stems rise above the water surface.

endemic: Having a native range restricted to a certain geographic area. For example, the Florida scrub jay (*Aphelocoma coerulescens*) is a bird that lives exclusively in Florida.

epiphyte: A plant that grows on another plant.

erosion: In aquatic management, the process by which soil or rock are removed from the shoreline of an aquatic ecosystem via rain and runoff, wind, and wildlife.

evergreen: A plant that retains green foliage throughout the year.

family: A taxonomic group of related plants or animals ranking above genus and below order.

floodplain: A generally flat and low-lying area of land adjacent to a river or stream, formed mainly of river sediments and subject to flooding. Broadly, this term can refer to any area of land subject to flooding by any source.

generalist: An organism able to utilize a wide variety of resources or adapt to a wide range of environments. Regarding butterflies, these are species that host their larvae on a broader variety of plant families.

genus: A taxonomic group of related plants or animals ranking above species and below family.

germination: The process by which an organism develops from a seed or spore.

hammock: An often elevated area of forest, frequently characterized by hardwood trees, that forms an ecological "island" in contrast to a surrounding ecosystem. In

Florida, this term tends to refer to an area of stands of hardwood or palm tree that are on higher ground adjacent to wetlands.

high-water mark: The highest level on land reached by rising water. In aquatic management, knowing the high-water mark helps us determine what species of plants we can place where based on those species' tolerance of inundation.

host plant: A plant that provides food or shelter for a particular organism. In this book, we use "host plant" to refer to a species of plant on which a certain butterfly species can lay its eggs.

hydric: Referring to a habitat that is permanently or seasonally saturated by water.

inundated: Flooded or submerged in water.

invasive: A plant that has been introduced to an area outside of its natural range (intentionally or otherwise) and disrupts native communities, sometimes to the extent of erasing native plant populations.

keystone: A species on which other species in an ecosystem largely depend, such that if it were removed, the ecosystem would change drastically.

lake: A relatively large body of water surrounded by land. There is no meaningful technical difference between a lake and a pond, but generally a lake is larger and deeper.

leaflet: A leaflike structure that, along with other leaflets, makes up a compound leaf.

littoral: In this book, we use "littoral" strictly in reference to an area with vegetation that is shallower than 2 meters but deeper than a meter during the rainy season. Outside of this book, it tends to mean any area where water meets land.

marsh: A type of ecosystem that is characterized by the dominance of herbaceous plants, rather than woody species, as well as being frequently or persistently inundated with water.

mesic: Referring to environments with a moderate amount of moisture.

monotypic: Containing only one species within a genus.

native: A species that occurs naturally in a particular area without human introduction.

nectar: A sugary, viscous fluid secreted by plants. While there are a handful of butterfly species that don't eat nectar, those that do value the substance for its nutrition and energy support in migration.

nitrogen fixer: An organism that converts atmospheric nitrogen into a form that is usable by plants. Incorporating nitrogen-fixing plants into the garden is a form of chemical-free soil enrichment.

ostiole: A small opening or pore, such as that of a fig through which pollinating wasps enter.

panicle: A type of inflorescence made of a loose, branching cluster of flowers.

perennial: A plant whose lifespan extends past two years.

phytoremediation: The process by which plants remove, transfer, stabilize, or destroy contaminants in soil, sediment, and water.

pollen: A powdery substance consisting of microscopic grains discharged from the male part of a flower or cone.

pond: A relatively small body of water surrounded by land.

prairie: A type of ecosystem characterized by the dominance of herbaceous species like grasses and wildflowers that is generally flat and relatively fertile.

propagation: The process by which new plants grow from various sources, such as seed or cuttings.

raceme: An unbranched inflorescence in which flowers are borne on short stalks along the main shoot.

rhizomes: Horizontal underground plant stems capable of producing the shoot and root systems of a new plant.

riparian: In this book, we use "riparian" to refer to the area generally between the high-water mark and the edge of the average depth of the water during the rainy season (which is generally no deeper than 5 centimeters).

river: A large, natural stream of water flowing in a channel to the sea, a lake, or another river.

rosette: A circular cluster of leaves, usually close to the ground and at the same height.

sandhill: A type of savanna-like ecosystem characterized by reliance on fire to maintain an understory of grasses and perennials, generally interspersed with pines or sand live oaks (*Quercus germinata*).

savanna: A type of ecosystem characterized by open grassland and dispersed trees. A savanna is a transitional ecosystem between a desert and a forest.

speciation: In this book, we use "speciation" to specifically refer to the process by which a group of plants within a species develops unique characteristics, typically based on geographical distribution, resulting in the creation of a new species.

species: The basic unit of biological classification, a group of living organisms capable of interbreeding.

stamen: The pollen-producing, male reproductive organ of a flower, typically consisting of a filament and an anther.

stream: A flowing body of water more or less forming a channel. While there are no strict, formal differences, a river, stream, and creek are generally the same type of water body but vary in size (largest to smallest in the aforementioned order).

swale: Generally, a somewhat depressed stretch of land with gently sloping sides that seasonally holds water.

swamp: A type of ecosystem that is permanently saturated with water and forested.

syconia: The fig inflorescence, a type of multiple fruit.

symbiotic: Involving interaction between two different organisms living in close association, typically to the advantage of both but occasionally harmful.

tannin: A bitter compound in some plants that binds to proteins, minerals, and more, resulting in insoluble compounds known for causing brownish colorations in water bodies.

tuber: A thickened underground part of a stem or rhizome.

upland: In this book, we use "upland" to refer to areas of higher elevation in comparison to riparian, emergent, and littoral zones of water bodies.

ACKNOWLEDGMENTS

We dedicate this book to the Holbrook family and their legacy. Without the direct support of Susan, Devin, Beth, Henry, and Arya, millions of plants and wildlife would not be here today—let alone this book!

Sean

There are many who have tirelessly supported and worked to help me be the scientist I am today and without whom I would be lost. To my mentors who taught me, my family who supported me, my friends who comforted me, and my staff who got it done—thank you. I could not have done it without you and will continue to do it with you until we succeed in saving Florida's ecosystems and restoring them to our communities.

Kendall

Simone Weil said, "Attention is the rarest and purest form of generosity." I have felt the truth of these words in the actions of my colleagues, friends, and family who have provided listening ears and keen words of wisdom and enthusiasm for this endeavor. Thank you for instilling, inspiring, and uplifting the sense of connectedness with the natural world that moves me to share its jewels.

We are deeply thankful for the image contributions of the friends and experts in the field who graciously provided their excellent photos to us for use in this book:

- Holly Greening (@hsgzoe on iNaturalist), a retired aquatic ecologist whose impressive work in photographing and identifying native flora and fauna shines.
- Clint Gibson (@cpgibson on iNaturalist), a conservation biologist and field researcher at the Florida Museum of Natural History who uplifts pollinators not only through his photography but also in his work on rare pollinator studies within the state's protected lands.
- Mark and Holly Salvato (@markhollysalvato on iNaturalist), who have together devoted a quarter of a century to the study and documentation of the Florida leafwing and Bartram's scrub hairstreak.
- Forest Hecker, environmentally focused outreach specialist and educator at University of Florida/Institute of Food and Agricultural Sciences Extension of Sarasota County, whose efforts in supporting the survival and reproduction of the atala butterfly cannot go unnoticed.
- Piper Cole, who spent innumerable hours creating beautiful reference graphics (as seen on page 18) for our team over the years; she also lent a tremendous hand in research and composition for this book in its early stages, for which we are immensely grateful.

We are infinitely appreciative of the team at Pineapple Press who passionately recognized the importance of this topic and have worked energetically and vigorously to support its presentation to the world.

Lastly, we would like to thank our readers. Without your interest and initiative, this long-hidden realm of habitat restoration would remain a mystery. We hope this book will contribute to a personal blooming toward stewardship and an informed, vivid sight on the world around you.

INDEX

Page references for photos are italicized.

ABOUT THE AUTHORS

Sean Patton is an aquatic biologist and the founder of Stocking Savvy, an environmental consulting company that specializes in the restoration of native plants and wildlife back into Florida's ecosystems. After being disillusioned by the traditional, costly, impermanent, and ineffective practices within the private sector of environmental management, he began researching alternative methods of restoring the state's landscapes to their most healthy and resilient forms. These efforts culminated in his practice of multimodal biological controls, a holistic approach to land and lake management. He currently serves as a professor of biodiversity at Ringling College of Art & Design, as well as the ecoflora coordinator at Marie Selby Botanical Gardens. If you'd like to get in contact with him, email stockingsavvy@gmail.com.

Kendall Southworth is a senior environmental consultant at Stocking Savvy as well as the company's manager. She has a bachelor's degree in environmental studies from New College of Florida. She began her habitat restoration journey at nineteen, when she joined the Stocking Savvy team as an intern. In the years since, she has aided in the design and installation of more than a hundred restoration projects across the state, helped produce the suncoast's largest environmentally oriented event in history, and works to educate the public on ecologically informed land and lake management practices. If you'd like to get in contact with her, email kendallsouthworth@gmail.com.